UP OR OUT
How to Get Promoted as the Army Draws Down

Wilson L. Walker
Master Sergeant, U.S. Army, Retired

IMPACT PUBLICATIONS
Manassas Park, VA

UP OR OUT
How to Get Promoted as the Army Draws Down

Copyright © 1993 by Wilson L. Walker

All rights reserved. Printed in the United States of America. No part of this book may be used or reproduced in any manner whatsoever without written permission of the publisher: IMPACT PUBLICATIONS, 9201-N Manassas Drive, Manassas Park, VA 22111, Tel. 703/361-7300.

Library of Congress Cataloguing-in-Publication Data

Walker, Wilson L., 1944-
    Up or out : how to get promoted as the Army draws down / Wilson L. Walker
        p.   cm.
    Includes index.
    ISBN 0-942710-91-6 : $13.95
    1. United States. Army—Vocational guidance. 2. United States. Army—Promotions. 3. United States. Army—Examinations
    I. Title.
    UB323.W28 1993
    355'.0023'73—dc20
                                               93-18015
                                                    CIP

CONTENTS

Acknowledgements vi

Preface vii

CHAPTER ONE
How to Prepare For Promotion 1
- Moving Up 2
- Self-Confidence and Faith 3
- Attitude and Motivation 3
- Follow Orders 5
- Be On Time 7
- Secure Performance Statements 8
- Be Your Best On Guard Duty 9
- Face-to-Face With Superiors 10
- Speak With Authority 12
- Test Your Military Knowledge 12
- Tips On Making Top Soldier on Guard Duty 15
- Care For Your Equipment 16
- Your Military Bearing 21
- Attitude 25
- Learn Your Job 25
- Score High On Your Physical Fitness Test 27
- Watch Your Weight 28
- Score 100 Percent on Your Skill Qualification Test 28
- Know Your Weapon 30
- Learn From Others 31
- Practice Soldier Care 33

CHAPTER TWO
How to Find Out If You Are Eligible 36
- Time in Service and Grade 36
- Physical Fitness Test 39
- Weapons Qualifications 41
- Height and Weight 41
- Promotion Points 42
- Security Clearances 42
- Promotion Recommendations 42
- Waivers 43
- Leadership Schooling 44
- Civilian Education 44
- Alcohol and Drug Prevention and Control Program 45

CHAPTER THREE
How to Get Promotion Points 46
- Skill Qualification Test 48
- Duty Performance 49
- Military Education 49
- Civilian Education 51
- Military Training 51
- Awards and Decorations 52
- Promotion Board 52

CHAPTER FOUR
How to Prepare For the Promotion Board 57
- Learn About Board Members 58
- Make a Study Guide 61
- Check Your Record 62
- Study Your Promotion Guide 63
- Prepare Your Uniform 63
- Request a Pre-Board 64

CHAPTER FIVE
How to Conduct Yourself During Board Procedures 66
- Dress in Accordance With AR670-1 66
- Become Knowledgeable About Board Members 68

Contents v

- Know Basic Soldiering 68
- Express Yourself Well 68
- Show Respect For the Board Members' Rank 69
- Depart Correctly 69

CHAPTER SIX
How to Prepare For
E-7, E-8 and E-9 Promotions 71
- Check Your Records 71
- Double Check Your NCOER 72
- Make a Photo 74
- Send a Letter to the Board President 74

CHAPTER SEVEN
Promotion to E-7, E-8 and E-9 76
- Eligibility For Promotion 76
- Written Communication 77
- Selection Board Results 78
- Effective Date of Rank 78
- Acceptance 79

Appendix 80
- How to Visualize 80
- How to Spit Shine Boots 85
- How to Set Up Your Wall Locker 91
- How to Sell Yourself 98
- Chain of Command 105
- Support Chain of Command 105
- Other VIP's in Chain of Command 106

Index 107

About the Author 110

Career Resources 111

ACKNOWLEDGEMENTS

My thanks to all the men and women in military uniforms, all over the world. This book is for you.

I want to thank my mother, Louise, who always told me *"You can do anything, if you truly want to."* Thanks to my wife, Gloria, who retired from the Army after Twenty years. Thank you, Wilson Jr., Carla, and Dawn (my children).

Thanks to all the soldiers that I have worked with through the years. Keep up the good work.

                                        Wilson L. Walker
                                    Master Sergeant, U.S. Army, Retired

PREFACE

Each person in the Army, at one time or another, has dreamed of being promoted. Some of the reasons are:

- More money for the soldier and his family
- A better or more rewarding job
- The desire to serve in a leadership position
- Recognition for job performance and rank
- More prestige

They also may have many other reasons relating to helping the soldier and his family.

During my twenty-one years in the Army, I've found that promotion was the number one subject among soldiers all over the world. I've seen countless numbers of soldiers miss getting promoted because they didn't have enough points or didn't prepare themselves for promotion. In the following pages I will show you:

- How to prepare yourself for promotion
- How to find out if you are eligible
- How to get promotion points
- How to prepare yourself for the promotion board
- How to conduct yourself during board procedures
- How to prepare for E-7, E-8, and E-9 promotions

This book will not only help you get promoted, but it will help you prepare your soldiers for promotions and other boards.

Chapter One

HOW TO PREPARE FOR PROMOTION

Getting promoted in today's Army is more challenging and competitive than ever before. Base closures and personnel draw downs mean an even more competitive environment for advancement within the Army of today and tomorrow.

If you plan to get ahead in the Army of the 1990s, you must be different from many of your peers who will not be as successful as you in getting promoted. You must both meet and exceed the expectations of those who can make a difference in your Army career. Above all, you must present your very best self to your boss and members of the promotion board.

That's my task in the following pages—to show you how to be a first-rate soldier and present your very best self to your boss and members of the promotion board. This involves everything from maintaining the proper military bearing to clearly communicating your accomplishments to your superiors. Your goal is to get promoted as rapidly as possible. It's a very simple goal requiring clear thinking, a positive attitude, hard work, perseverance, and certain key communication skills.

If you follow the advice in this book, you can indeed move up as the Army draws down. You'll get ahead in today's Army. Best of all, you'll both feel good about yourself and your contributions to others.

MOVING UP

You must know yourself one hundred percent. Before you can sell yourself to your boss or members of the promotion board, or any other board, you have to know yourself because:

- You will know your strong and weak points.
- You can get help, if you need help.
- You can make plans for your promotion.
- You can spot check your progress.

Once you recognize your weak points, you can begin working on them. When the time comes for you to appear at the promotion board, you will have it all together; you will need to maintain your strong points and work on making your weak points stronger. If you need help, get it. You can do this by:

- Reading about what you need help with.
- Asking someone else for help.
- Watching other people do what you would like to do.
- Looking at movies that motivate you.

There are books on just about anything you want to know. You have to find the books; call someone at the bookstore about the books you are looking for. No bookstore will have all of the books.

You can ask someone to help you with your weak points and, at the same time, you may be able to help them with something. Find out how they got promoted or how they were able to make 300 on the PT Test. Watch other people do things you would like to do. Go to a pre-board and watch others perform; talk to them about it. People as a whole like to help each other, especially when it has something to do with what they are good at doing. You can look at movies that motivate you; my motivation movie is *Rocky*.

SELF-CONFIDENCE AND FAITH

You must have absolute faith and confidence in yourself; know your self-worth and your desire to get ahead. Make yourself number one. You have to be and stay number one; let no one come before you. I don't think anyone on this earth loves his mother, wife or children as much as I love mine yet, I love me the most, next to God that is. If you don't love yourself, how can you love another? If you don't love yourself, you are not going to care about getting promoted or having the things that will make your life more rewarding. The way you feel about others and yourself is your attitude, which is nothing more than the way you think and feel about something. Don't let anything hold you back; always prepare yourself, look ahead, make plans and stick to them. Talk to your loved ones about your plans and your promotion. I have seen soldiers that didn't go to a leadership school because they:

- Didn't have the money.
- Didn't want to leave their wife and kids behind.
- Had no confidence in themselves.
- Were overweight and couldn't pass the PT Test.

No one wants to leave their loved ones behind, but there are times when you will have to, whether it's when you go to the field or TDY. The thing to do is talk to your loved ones about it and make plans for your absences.

Some soldiers have a lack of confidence in themselves when it's time to go to a leadership school. This is why you must know yourself and be prepared. When the time comes you will have no hang-ups. Also, this is true when the time comes to pass the PT Test or with keeping your weight at an acceptable level. Love and make yourself number one; everything else will fall into place.

ATTITUDE AND MOTIVATION

Develop and keep a positive attitude. How you feel about yourself and your success depends a great deal upon your attitude. We said earlier that the attitude is nothing more that the *"Way You Think"* about yourself, your family or your job. When you change your attitude, you change your life because, in the process, you change the way you feel about things. Attitude is the most important part of motivation. So, what is motivation? Motivation is the *"Reason for Action,"* why we do the things we do. Let's put the two together to see how they work hand-in-hand. Now don't forget! Attitude is the way we think; motivation is the reason for action. Let's say that you have a PT Test coming up in two months and you want to make 300 on the test because it will give you more promotion points. First you must get a positive attitude about the test, which is the way you think. You can do only what you think you can do. You must prepare. Now, you think you can make 300 on the test. Next, there's the motivation or the reason for action. Why do you want to make 300 on the PT Test? You want to make 300 on the test so you can get more promotion points. From motivation there are always benefits and the benefits in this case are:

- Move promotion points.
- Better physical condition.
- Promotion to the next highest grade.
- Being awarded the Army's Physical Fitness Patch.
- Being in the Top Ten Percent in the Army.

We can say that Attitude + Motivation = Benefits. Let's look at another example. Say you want a new car. The way you feel about it is your attitude. Next comes motivation, the reason for action, the reason you want the car. After you put the two together, you will get the benefits and they could be:

- Saving money on repairs.
- More confidence in family security.
- Pride in ownership.

Promotions don't happen overnight. You have to be motivated and stay motivated. When you know your reason for action, you will be able to maintain action. We do the things we do for some type of benefits. Motivation is more than feeling; it is knowing your benefits and taking action to achieve those benefits. To maintain your motivation daily, begin now and read, listen and observe a little each day. Maintain self confidence and a productive attitude. Pat yourself on the back each day for something you accomplish, whether big or small. Don't wait for others to do it because you will always accomplish something others don't know about. There is nothing more motivating than accomplishment.

The key to personal and professional success isn't merely getting motivated; it's maintaining that motivation. You should start preparing for promotion as soon as you enlist in the Army or as soon as you think of enlisting. Visualize becoming a Command Sergeant Major or the Sergeant Major of the Army. This will be your long range goal and you will have to make E-5 and all the other ranks before you get to Sergeant Major. As soon as you make one rank, keep visualizing eventually making the Sergeant Major rank. Read How to Visualize in the Appendix at the end of this book. Getting promoted from E-1 to E-4 is not as hard as getting promoted from E-5 to E-9; it takes more to go from E-4 to E-6 because of the promotion board. At the time you are trying to make E-1 through E-6 you will be preparing your records and yourself for promotion to E-7 through E-9. Some of the things that may help you make E-1 through E-4 are:

- Doing as you are told.
- Be at your appointed place of duty on time.
- Get your performance and promotion counsel statements.
- Be the top soldier on guard duty.
- Follow tips on making top soldier on guard duty.
- Take care of all your equipment.
- Meet the standard for military bearing.
- Learn all you can about your job.
- Score 290 or more on your Physical Training Test.
- Never gain more than one pound.
- Score 100 percent on your SQT.
- Become an Inspector or Evaluator.
- Become an expert with your weapon.
- Learn from others.

FOLLOW ORDERS

Do as you are told. Many soldiers get into trouble every day for not doing what they are told to do. Remember, it's very important to do as you are told, when you are told—especially if told by an officer or non-commissioned officer. When an officer or non-commissioned officer tells you to do something, it is an order. You don't have to be told, *"That Is An Order"* or *"I Order You."* When an officer or non-commissioned officer tells you to do something, never tell them:

- You don't have the time.
- Talk to your boss.
- You were told to do something else.
- That's not your job.
- You are off duty.
- You have to talk to your boss, first.

All soldiers take orders from officers and non-commissioned officers that out rank them or hold a higher position. This applies to all soldiers, but happens more often to the E-4's and below. Most orders are given to the E-4's and below by a non-commissioned officer or an officer. The only times E-4's and below will give orders are:

- They are in charge.
- They are an acting non-commissioned officer.
- They are relaying an order from an officer or non-commissioned officer.
- They out rank the others and have the need to issue orders.

Most of the time if an officer or non-commissioned officer is not around, they will leave the next highest ranking soldier in charge. This is done by rank. If one or more soldiers have the same rank, the one with the most time in grade will be in charge. If the time in grade is the same, the one that has been in the Army the longest will be in charge and, if that is the same, the oldest will be in charge. An E-4 or below will at one time or another be in charge; they may be left in charge when the boss is away or just be in charge of that section. Most of the time, this will be an E-4 or E-3. There are not many times that you will see a PVT, E-1 or E-2 in charge. E-4's, as Acting Sergeants, have the same power that other Sergeants have. Most Acting Sergeants have been to the promotion board and are waiting to get promoted. Some have been to leadership school or they have leadership qualities. Sometimes, an officer or non-commissioned officer will have one soldier tell another soldier something; this is relaying an order. The soldier will restate, to the other soldier, the order given him. There are times when an E-4 or below will have to take charge and he will tell the others what to do. If something has to be done, and the boss is not around, take charge and get it done. Let your boss know what you did. When an officer or NCO tells you to do something and you tell him to talk to your boss, you have made two very big mistakes and each can cause you to lose money and a stripe, or you can be put out of the Army. You disobeyed a lawful order and you were disrespectful to the officer or NCO. Always take the order; do as you are told. You are a soldier twenty-four hours a day; never say you don't have the time. If you are off duty and are told to do something, do it. There will be many times in the Army that you, as a soldier, will be told to do something and you won't know the whole deal or see the whole picture, but stay out of trouble; do it!

BE ON TIME

Always be at your appointed place of duty on time. You will be shocked at the number of soldiers that get into trouble because they were not at their place of duty, or were not there on time. Your appointed place of duty is where you were told to be and it may not be your day to day duty. It could be:

- Physical training or work formation
- Sick call or the hospital
- Firing range or field site
- Detail with other soldiers or another company
- First Sergeant's or Commander's office
- Temporary duty assignment

Any place you are told to be is your appointed place of duty. Once you are told to be there, that is an order. Always be at your place of duty fifteen to thirty minutes before time. That way, you won't be late. Sometimes you will have more than one place of duty during a given day. You may have a doctor's appointment at the hospital and that may be your appointed place of duty for 07:30. However, you may have to make a 06:00 PT formation at your company (PT formations are also used for accountability). After the formation you will go to your 07:30 appointment at the hospital. You may be told to be at another company at 08:00 the next morning. Be sure to check to see if you have a formation, or call in to your company. It's best to double check by calling in. When you are sent to school or TDY, always be on time because going to school or TDY is your place of duty. Always be on time everywhere you go. Be at the right place at the right time. Never be late when you are told to report to the First Sergeant's or the Commander's office. Who knows? You may be going there to get promoted. Anytime you have guard, CQ or any other kind of duty, please be on time, in the right uniform and have whatever equipment you may need to perform that duty. If you change your duty, be sure to let whoever is in charge of the duty roster know that you are changing duty. Most of the time it doesn't have to go any farther than your boss, the Platoon Sergeant or the First Sergeant.

SECURE PERFORMANCE STATEMENTS

Get your performance and promotion counsel statement. All of the E-4's and below should get a monthly performance and promotion counseling statement. If you are an E-4 or below and do not get one, ask for one because:

- You'll see how your boss views your job performance.
- You become aware of your strong and weak areas.
- It helps you to prepare for your SQT.
- It motivates you to do better.

Once you get your performance statement, read it so you can see how your boss feels about your job performance. You may think you are doing good, but your boss states problem areas; on the other hand, your boss may think you are doing better than you think. Always get one, read it and keep it. Job performance will get you promoted faster than anything, from E-1 to E-9. Always do your best and more. Check out your strong and weak points. Try to improve each because that also will help you when the time comes for your SQT. If you want to get promoted and be a good soldier, your statements will motivate you. Ask for a copy of your statement each month and keep them. Most of the time your boss will keep them, but ask him to make a copy for you. Keep asking until you get one and keep them. The main reason you want to hold on to your statements is, one day you may have to talk to someone about getting promoted and your statements will show the good job you have been doing. If for some reason you get into some kind of trouble, most commanders will ask about your job performance. If you are ever faced with an Article 15, you have the right to show your statements to your commander. You want to always get good statements because if you get bad ones your boss can use them against you. You want to make sure your boss has your job book with him each day. If he doesn't have one for you, don't sit around and wait, try to get one yourself and give it to him. If you can get two, you can keep one for yourself. Whenever you do something that is in the job book be sure and tell your boss about it so he can fill in the date, and you do the same if you have one. When the time comes for your SQT, the job book could be a big help. So get one, or make sure your boss has one for you. Next come your promotion counseling statements, they are just as important as your performance statements and your job book, if not more, because:

- You get a feel about your promotion.
- You know what you need to do or not do.
- You can check your progress.
- You have proof for complaints.

Just as your performance statements, when you get your promotion counseling statements read them and keep them. If you don't get a promotion counseling statement, have your boss comment about your promotion on your performance statement each month. If you do get one, you will get a sense about how you are doing and you will know what you need to do in order to get promoted, or what not to do. You can also check on your progress each month. If you are doing things right and still not getting promoted, you can take in the performance statements and file a complaint.

BE YOUR BEST ON GUARD DUTY

Be the top soldier on guard duty. Guard duty should be your training aid for future promotion boards because:

- You start training early.
- You come face-to-face with your superiors.
- You stand inspection.
- You learn to speak with authority.
- You test your military knowledge.
- You get a chance to sell yourself.

You start training early for promotion boards just by pulling guard duty. Most of the time this will be at:

- Your basic training unit.
- Your advance individual training unit.
- Your permanent party station.

Each time you pull guard duty, you will learn more about:

- Yourself
- Your superiors
- Military subjects
- Your weapon
- Your peers

Again, you must know yourself, know what to do and how well you can or cannot do something. You need to know all you can about your superiors, weapons and peers. Knowing these things will also help you when the time comes for you to go to the promotion board. Each time you pull guard duty, be the best. Look at it as a training aid; it will help you in the long run.

FACE-TO-FACE WITH SUPERIORS

You come face-to-face with your superiors when you pull guard duty and at the promotion board. You are closer on guard duty than you are at the promotion board. Guard duty is where you will learn how to face your superiors. Once you can face them on guard, facing others at the promotion board will be easy. On guard duty always look your superiors in the eye, not eyes. Look in the right eye or left eye. You do this because you can focus better on one point than you can on one thing when you are close up. While on guard duty, always look into the eye of the Officer Of the Day, or the Sergeant Of The Guard when they are inspecting you. At the promotion board, you are farther away, you will look into the eyes because it's harder to look at a spot far away. Or you will look at the board member's face or something near his face. They will not be able to tell. Whenever you talk to anyone close up, look into the eye. Whenever you are far away, look at the face or something near the face.

How to Prepare For Promotion

You stand an inspection on guard duty and you will be inspected at the promotion board. On guard duty, you will be close to your superiors. They will be able to see things close up like:

- Lint and short strings on the uniform
- Mustaches and shaves
- Arrangement of your gear
- Boots or shoes
- Weapon and other equipment

When you report to the president of the promotion board, you may be within three to five feet of the president and the members of the board. They will be able to see:

- How well your uniform fits.
- Arrangement of Medals and Ribbons.
- Hair length and haircuts.
- Military bearing.

If you stand guard mount with your weapon, be sure that you know your weapon number. You should know it as well as you know your name and social security number. Clean your weapon very well and have two people inspect it for you before you go on guard mount. Know all the parts of your weapon and everything else there is to know about it. When you are being inspected on guard mount, as soon as the Officer Of The Day or The Sergeant Of The Guard comes in front of you and faces you, bring your weapon to inspection arms and hold it very loosely. The Officer Of The Day, or whoever is doing the inspection, may or may not take your weapon to inspect it. If they reach for it, let it go. The best way to tell is to look into the eye and, at the same time, glance at the shoulders. Once you see their eyes or shoulder move, let the weapon go, If they are not reaching for it they will. The more you practice doing this, the better you will get. You can practice:

- When you have guard duty.
- Anytime you clean your weapon.
- When you are in the field.
- Whenever you have it out of the Arms Room.

Never practice this at the firing range!

SPEAK WITH AUTHORITY

You should learn to speak with authority because:

- It helps prevent nervousness.
- It puts you out in front of your peers.
- It keeps you on the inspectors' minds.

When the inspector on guard mount is facing you and asking you questions, sound off loud and clear. Let him and your peers know that you are the top soldier on guard duty today. This not only lets them know that you are the best, it keeps you from being too nervous. The Officer Of The Day, or whoever is the inspector for the guards, will not forget you when it comes time for him to pick the top soldier for guard that day. After he inspects all the guards, he picks one for top soldier. He may not remember who had the most correct answers or who was dressed the best, but he will remember who sounded off. So, let everyone know that you are the best.

TEST YOUR MILITARY KNOWLEDGE

You test your military knowledge each time you pull guard duty. You may be asked questions about yourself, the state you are from, or there may be questions about sports; but you can be sure there will be questions about the military. It would help if you knew:

- What questions will be asked.
- Who will be the Duty Officer or The Sergeant of The Guard.
- Who will be your competition.
- What to study.

The first thing you need to do when you start pulling guard duty is make up a study guide. To do this, get a notebook and find out who inspects the guards. It may be the Duty Officer, The Sergeant of The Guard or, in some places, The Staff Duty Officer or The Staff Duty Non-Commissioned Officer. Whoever it is, find out who is inspecting.

How to Prepare For Promotion 13

Find out where their roster is posted. A copy should be on the board with the CQ and other rosters; if not, find out where and get a copy or write down the names from the roster. You will have to keep doing this until you get the names of all the inspectors. Keep checking and once a new inspector comes in, write down that name also. After you get all the names, list them on the first page of your notebook in order by rank. You may need about three pages for each inspector. Next to the name, write down the page numbers for each inspector. When this is done, it should look something like this:

NAMES	PAGE
1LT Adams	1-3
1LT Benson	4-6
2LT Mason	7-9
CW2 Page	10-12
MSG House	13-15

Once you get the names of the inspectors and the page numbers on the first page or the table of contents, you will only need the subjects. This will not be the same for each inspector. Some will have the same subjects but not the same questions. One may have a subject, the others have never used, but each will have his own subjects and questions because:

- They may be his favorites.
- They are easy to remember.
- The inspector knows the subject well.
- They don't take time to prepare new ones.

Each time you pull guard duty, after guard mount, get your book and turn to the inspector's name and write down the subject, the question asked, and the answer. If you were asked about other subjects, write them down with the questions and answers. It may look like this:

1LT ADAM

Weapons:

Q—What is your weapon number?

A—Sir, my weapon number is _____.

Chain of Command:

Q—Who is your commander?

A—Sir, my commander is _____.

For each inspector, you will make up a list of questions and answers asked about the subjects. You don't have to wait until you have guard duty to get your subjects, questions and answers. Talk to the other soldiers and ask them what was asked by the inspectors. That way, you will get your guide full faster. Work on your guide whenever you can. This may be:

- On guard duty.
- CQ runner.
- When you go to the field.
- When you are off duty.
- Anytime you have nothing to do.

You get a chance to sell yourself when you pull guard duty and when you go before the promotion board. You have to be able to do this well and better than your peers. If you are going to be the top soldier on guard mount, or get the maximum amount of points and a recommendation from the promotion board, you can sell yourself well if:

- You know the subject that's covered well.
- You have confidence in yourself.
- You know the inspectors or board members.

Know your subjects well. For guard duty, make your study guide. I will tell you how to get one for promotion when I talk more about the promotion board preparations. You must have confidence in yourself and know what to do. Knowing your subjects well will give you confidence in yourself. You should also know who the inspector or board members are.

TIPS ON MAKING TOP SOLDIER ON GUARD DUTY

All you have to do to become Top Soldier on Guard Duty is be the best. Always try to stand in the back of the formation, or as far from the front as possible, because:

- You can check out your peers.
- You may hear some of the questions.
- You can make plans for your turn.
- You get rid of tension.

Always check out your peers. Some of them will try to make top soldier; know who they are. Know their weak and strong points. If you are in the back of the formation, or one or two ranks away, you may be able to hear what questions the inspector is asking. He may or may not ask you the same questions. If you have made your study guide, you will have an idea of what questions he may ask because:

- Most inspectors use the same questions.
- Some didn't take the time to get new questions.
- They may not be prepared for duty.

Like you, the inspector's primary job is not guard duty. He may have been tied up in his primary job so much that he didn't take time to think about what subjects or questions he would cover. For this reason, he may have to use the same questions that he used before. This is one reason you should make a study guide for guard duty. As the inspector gets closer to you, you can make plans as to what you will do when he is in front of you. For example, you might sound off louder than the other soldiers sounded off. Also, you will have more time to relax and relieve tension before it is your time for inspection.

CARE FOR YOUR EQUIPMENT

Take care of all your equipment. You are responsible for it. Once you are in the military, you will be responsible for all types of equipment. Some of the equipment may be:

- Your assigned weapon
- Your field equipment
- Your basic issue
- Your vehicle and tools
- Your equipment storage area
- Your assigned room and wall locker

Your assigned weapon is the most important equipment you will ever be assigned, because one day it may save your life; take care of it. All Army equipment comes with a book that tells you how to take care of it. Be sure to get the book. It may be a Technical Manual (TM), Field Manual (FM), Lub Order (LO) or Standard Operation Procedures (SOP). Be sure you know which book is for your equipment and use it. Always remember you can't go wrong with the book.

Take care of your equipment and it will take care of you. Your assigned weapon may be:

- An M16A1, M16A2 with Bayonet
- An M203, 45 or 9mm

Along with your basic weapon, you may be assigned other weapons such as:

- 50 Cal Machine Gun
- M60 Machine Gun
- M72A2 LAW (Light Antitank Weapon)
- Hand Grenades
- M18A1 (Claymore Mine)

Also, you could be assigned part of a weapon system. These systems are very important to the Army's Defense. Some of these systems are:

How to Prepare For Promotion 17

- Armory (Tanks)
- Aviation (Planes)
- Missile Firing System (Stinger)
- Chemical Weapons Teams
- Air Defense Artillery (Missiles)

All of your equipment may not be combat equipment. If it is not combat equipment, it will be combat related. Also, there are all types of repairmen equipment, such as:

- Telephone repair
- Field Radio repair
- Construction equipment repair
- Light wheel vehicle repair
- Abrams and other tanks repair
- Fuel and electrical systems repair
- Utility airplane repair
- Helicopter repair

There are many more and there is other equipment that is just as important. Some of this equipment is associated with:

- Infantry
- Combat Engineers
- Field Artillery
- Armory
- Audio Visual
- Communication
- Supply
- Transportation
- Medical
- Food Service
- Military Police
- Bands
- Bridge Crew
- Ground Surveillance Radar
- Fire Fighter

And so much more. Almost everything you can find outside of the Army, you can find in the Army, only more. The main thing for you to remember is to take care of your equipment. Get the TM's, FM's, LO's and SOP's. All of this will show you how to take care of your equipment.

Your field equipment is very important because this is the equipment you will take to war with you. The amount of field equipment you are issued will depend on what type of job you have. Some equipment is:

- TA-50 field gear
- NBC equipment
- Your basic weapon and other weapons
- Transportation
- Mess Gear (MKT and Mess Kits)
- Tents and sleeping bags
- Alice packs
- BDU's, field jacket and personal gear

and any other equipment that you take to the field or that was made for the field.

Your basic issue will stay with you for as long as you are in the Army. In most cases you will keep it when you get out of the Army. You will have to repair or replace some of the equipment. You will get money each month to care for or replace your equipment. Some of your basic issue is:

- The class A uniform.
- The battle dress uniform and cap.
- Your shoes and boots.
- Underwear and socks.
- Shirts and blouses.
- Laundry bags and field cap.
- Overcoat and field jacket.
- Money for females' personal gear.

You will always take this equipment with you when you go:

- To another duty station.
- To a leadership or job related school.
- To temporary duty station.

How to Prepare For Promotion 19

You may get lucky and get stationed where you wear only civilian clothes. If so, you will be given money to buy the clothes, but you will take your basic issue with you also. Always repair or replace your basic issue when needed.

Your vehicle and tools should be cared for the same as your other equipment. Again, get the TM's for your equipment, so that you will know how to take care of it and when it should be serviced. Keep your tools clean and turn them in if they are unserviceable. Do your inventory when it is due and keep your tools locked up. Soldiers who have tools pay more money for lost tools than any other equipment. Take care of your tools and keep your money. Some of the vehicles that you may have are:

- Ambulances
- Trucks
- Tanks
- Hawk missile loader transporter
- Staff, MP's and TMP cars
- Equipment vans
- Tools vans
- Communication vans and trucks
- Dump trucks and many other types of vehicles

Some of your tools may be:

- Mechanic tools
- Plumber tools
- Field radio tools
- Missile maintenance tools
- Electrical tools
- Aviation repairmen tools

Your equipment storage area could be:

- A tool room
- A storage van
- An ammo can or box with locking device
- An old wall locker
- A truck with storage equipment on back

- The motor pool or track park
- Warehouses and many others

Wherever you keep your equipment stored, keep it neat and clean. Make sure the equipment is organized, and that all the unserviceable equipment is tagged for turn in. Also, turn in any excess equipment you may have in your storage areas. One thing you need to know is that some high ranking officers and NCO's will look into your storage areas. If yours are in bad shape, do you think they will remember it when you are put in for a promotion or when you go before the promotion board? You should always clean and store your field equipment as soon as you return from the field. Never put it off; you may need it the next day.

Your assigned room and wall locker can make or break you because:

- It shows what kind of person you are.
- It can be a plus for your promotion.
- It can get you into a one or two man room.
- It keeps others out of your room.

The soldier that is neat and clean keeps his room the same way. Look at the soldiers around you and then look at their rooms. How does your room look? Always have it ready for the Sergeant Major of the Army or the Chief of Staff to see. You never know when they may come into your room. Just think what would happen to your promotion if one of them came and your room was not up to par. Don't worry, nine times out of ten they won't say anything to you; but they may say something to your Sergeant Major, who will tell your First Sergeant. If you keep your room clean, it can be a plus for your promotion. As soon as you are assigned to a room, ask for the room's SOP. If there is not an SOP, make one up for your room. If you have a room with another soldier, get with him and make up a room SOP. If you outrank him by grade or time in grade, you are the room commander. Have something for each soldier to do in the room each day. If you are in a three or four man room, keeping your area neat and clean may get you into a one or two man room. Once you get your room neat, clean and set up by the room's SOP, the other soldiers will not come into your room because:

- You won't let them.
- They don't want to help clean up.
- They feel out of place.
- They don't soldier as hard as you.

When you open your wall locker, it's like opening the door to your room because:

- It shows what kind of soldier you are.
- It can be a plus for your promotion.

A soldier that has a neat and clean wall locker and room will, more likely, be neat and clean every day. Get the SOP and set up your wall locker by it. Again, if there is not one, make up one; but make sure the other rooms are the same. If there is not an SOP, you can always ask your boss for one. If he cares he will get you one, because if you don't do your best it will make him look bad. In the Appendix you will learn "How to set up your wall locker" to pass all inspections.

YOUR MILITARY BEARING

Set the standard for military bearing. Military bearing is the way you carry yourself as a soldier, on or off duty, state side or overseas. Your standards for the military should be very high, whether you are a Private or a Sergeant Major. The way you sit, talk and walk are all part of your military bearing. The military bearing that is most noticeable by your superiors and other soldiers is:

- The way you wear your uniform.
- How well groomed you are.
- Your attitude toward other soldiers and officers.

The way you wear your uniform tells a great deal about your personality. Army Regulation 670-1 will explain:

- How to wear the uniform.
- When you can wear the uniform.
- What to wear with the uniform.

There are many types of uniforms; in the Army some are:

- The Class A Uniform
- The Dress Blue Uniform
- The Mess Uniform
- The Class B Uniform
- The Field Dress Uniform
- The Cold Weather Uniform
- The Battle Dress Uniform
- The Physical Training Uniform
- The Aviation Uniform
- The NBC Uniform (suit)
- The Mechanic Uniform (coveralls)
- The Medical Uniform (whites)
- The Food Service Uniform (whites)
- The Military Police Uniform
- The Military Intelligence Uniform
- The Maternity Uniform

The main dress uniform for male soldiers is the Class A Uniform, and the Classic Green Service Uniform for female soldiers. Some of the other dress uniforms are:

- The Dress Blues
- The Class A Uniform With White Shirt
- The Mess "Dining" Uniform
- The Class B Uniform

The main duty uniform is the Battle Dress Uniform (BDU).
There are other work uniforms, such as:

- The Flight Uniform
- The Mechanics Coveralls
- The Medical Uniform (Whites)
- The Food Service Uniform (Whites)
- The Military Intelligence Uniform
 (Depending on the duty)

There are other jobs where you can wear the BDU's with another uniform. Some of these uniforms are:

- The Cold Weather Uniform
- The NBC Suit
- The Mechanics Coveralls

Because of their jobs, some soldiers have more than one type of duty uniform. Some of these are:

- The Food Service Uniform
- The Military Intelligence Uniform
- The NBC Personnel Uniform
- The Medical Uniform

Your uniform is you! How much self-discipline do you have? Do you have pride in yourself? How do you feel about your appearance? You can look at another soldier and figure out the answers to these questions by the way he or she dresses and carries him or herself. Other soldiers can do the same thing by looking at you. Your uniform should always be clean, pressed and fit properly. Army Regulation 670-1 will tell you all about your uniform. There are other accessories that you can wear with your uniform. Some of these are:

- The Head Gear
- The Foot Gear
- Other Accessories

The head gear and uniforms may be different because of the job involved. Some of the head gear is:

- The Garrison Cap
- The Service Cap
- The BDU Cap
- The Drill Sergeant Hat
- The Green, Black, Red and Other Berets
- The Helmet
- The Black PT Cap and Many Others

Also, there are different types of footwear. Be sure to check AR 670-1 to see when you can wear the different footwear and with which uniform. Some of the footwear is:

- The Combat Boots
- The Jungle Boots
- The Cold Weather Boots
- The NBC Boots
- The Low Quarters
- The Pumps
- The Physical Training Shoes
- The Safety Shoes
- The Over Shoes

You must care for your footwear as well as your other equipment. If your PT shoes are wearing out, have them repaired or replaced. Do the same with your boots and low quarters. You will be able to turn in and have some of your footwear replaced. Be sure to keep your boots and dress shoes shined. Read "How To Spit Shine Boots" in the Appendix of this book. Some of the other accessories for the uniform are:

- The Special Skills Badges
- The Marksmanship Badge
- The Identification Badge
- The Rank Insignia
- The Brassards
- The Name and U. S. Army Tapes
- The Belts
- The Black All Weather Coat
- The Field Jacket
- The Gloves
- The Sweater
- The Windbreaker
- The Medals
- The Awards and much more, all found in AR 670-1

Set your standards very high for military bearing. Always be the best or among the best. Be sure to get and read AR 670-1. Try to get one that you can keep because you will use this regulation more than some of the others.

How well groomed are you? Your uniform may be clean and pressed, and your boots or shoes may be highly shined, but:

- Do you need a haircut?
- Did you shave this morning?
- How many rings are you wearing?
- How long is your mustache?
- How long are your side burns?
- What type of hairstyle do you have?
- What color are your fingernails?
- What color lipstick are you wearing?
- What color are your hairpins?
- When can you wear a wig?
- Can you wear a cross around your neck?
- Can you bleach your hair?
- Are your gold chains showing?
- Are you wearing too much make-up?

All of these questions are connected with your being well groomed. You can find the answers to these questions in AR 670-1. If you know the answer, it's good to double check because the other soldiers will always check you out. Some soldiers like to let others know how much they know about military regulations.

ATTITUDE

Your attitude toward other soldiers and officers should always be the best. Always show respect for others and treat them the way you would like to be treated. Don't believe in rumors. Find out what is going on, know where to look for help, and whom to talk to when you need help. Remember, attitude is the way you feel about something. Try to have a good attitude about other soldiers and officers.

LEARN YOUR JOB

Learn all you can about your job. Learn all you can about your job, as soon as you can, because it is the start of the promotion process. It can help you:

- Get 100 percent on your SQT.
- Get more promotion points.

- Become an inspector or evaluator.
- Become an instructor.

You will be given a Skill Qualification Test to see how well you know your job and how well you know it compared to the other soldiers in your grade and MOS. If you learn all you can about your job, you will do much better on the test. There are books that will tell you how to:

- Operate the equipment
- Do the job
- Service or fix the equipment
- Order parts for the equipment

Find out how many books there are for your job. When you get your SQT notice it will have a list of books you will need for the test. You can get these and other books about your job by asking:

- Your Supervisor
- Your Platoon Sergeant
- Your Warrant Officer
- Your TAMM Clerk
- Counselor at the MOS Library

See if you can get a set of books to keep. When you get them, read them. You can read about your job:

- On duty, during breaks
- On Guard Duty or when you have CQ
- On weekends and during FTX's

You will not be able to keep books that are classified. you will be able to read them and lock them up when you are finished.

SCORE HIGH ON YOUR PHYSICAL FITNESS TEST

Score 290 or more on your Physical Fitness Test. The Army Physical Fitness Test is a test designed to measure your strength, endurance and cardio-respiratory efficiency. This test consists of:

- Push-ups
- Sit-up
- Two mile run

The push-ups measure the strength and endurance of your arms, shoulders and the chest muscles and are performed for two minutes. The sit-ups measure the strength of your abdominal and flexor muscles and are performed for two minutes. The two mile run measures your cardio-respiratory fitness which is your heart, lungs and blood vessels. The run will also be timed depending on age and sex. All soldiers will take the Army's Physical Fitness Test at least twice a year. The test should be given at least four months apart. The test should not be given on a Monday or after a three or four day weekend or a holiday. You may take more than two Army Physical Fitness Tests during a given year because of:

- Reenlistments
- Promotions
- Schools (Military)
- Command or other inspections

Score 290 or more on your physical fitness test so that:

- You will be more physically fit.
- You will be in the Army's Top Ten Percent bracket.
- You will be eligible to wear the Army's Physical Fitness Patch.
- You will get promotion points.
- You may receive a reward from higher headquarters.

WATCH YOUR WEIGHT

Never gain more than one pound because:

- It can slow down your performance.
- It can cause your heart to work harder.
- It can cause your uniform to fit improperly.
- It can cause you to get put out of the Army.
- It can slow down your motivation.

It doesn't take much to gain or lose a pound. Keep an eye on your weight; don't let your weight keep you from getting promoted. You cannot get promoted or go before the promotion board if you are overweight. Being one or two pounds overweight can make you look bad in your uniform. When you go to the promotion board, you will be graded on your appearance. If you are boarded and become overweight after you are put on a standing list, you will not be promoted. Just gaining a pound can cause you to lose points from the points you already have because you may not do well on your physical fitness test before the next recomputation.

SCORE 100 PERCENT ON YOUR SKILL QUALIFICATION TEST

Scoring 100 percent on your skill qualification test will:

- Help you get a good assignment.
- Help you get ahead of your peers.

When your branch at the Department of the Army has to select someone for a special assignment, they select the rank needed and look at the SQT scores, because scores will show how much the soldier knows about his job. Some of the assignments are:

- Instructors
- Schools for the next higher grade
- Assignments to another country
- Selections for special duty
- Teachers at a college or high school

The important thing to remember about your Skill Qualification Test is that it is generally used as a method of evaluating your qualification to the next higher grade. If you and other soldiers were recommended for promotion (E-1 through E-4), and there were not enough stripes for each of you and your time in grade and service were about the same, nine times out of ten the soldier with the best Skill Qualification Test score will get the stripe. E-1 through E-4 do not need promotion points to make the cut-off score unless there is a board for the E-4's, and in most places there is not.

Becoming an Inspector or Evaluator is very rewarding because:

- You get more recognition.
- You are among the top soldiers in your MOS.
- You may get more money (TDY).
- You will learn more about your job.
- You get better assignments.

When you learn your job well and do well on your SQT, you may be picked for an Inspector or Evaluator. You may be part of:

- The AGI Team
- The MAIT Team
- The CMMI Team
- The CI Team

You will get more recognition from high ranking NCO's and Officers which could be very helpful for your promotion and future assignments. You will be among the top soldiers in your MOS. As an inspector or evaluator, you may get more money if you go on a temporary duty assignment. You will learn more about your job and you will:

- Do your job more.
- See how others do the job.
- Learn more about changes.
- Be up to date.

When you change duty station, you could get a better assignment or become an inspector or evaluator. If not, you will still know and learn more than you did before; and may get 100 percent on your SQT.

Becoming an instructor will help you learn more about your job and, at the same time, you will be teaching others. If you become an instructor, some of the soldiers you teach may work for you someday or, someday, you may work for them. Either way, they will know that you are good at your job and this can help you get promoted. Job performance is very important. Be the best or among the best!

KNOW YOUR WEAPON

Become an expert with your weapon. Your basic weapon will be your main protection if you are called to war. Become an expert with your weapon and you may survive. The way to become an expert is to:

- Take care of your weapon.
- Zero your weapon (M16A1 & A2, M203, etc.).
- Become an expert.
- Continue to train.

Take care of your weapon. Keep it clean and serviceable. When you clean your weapon, look at it very closely; check each part for damages. If there are any damages, report it to your supervisor and record it on a DA Form 2404. Always take the time to do a good job when you clean your weapon. You should clean it:

- At least once a month.
- When you return from field training.
- Before and after the firing range.
- Before you go on leave or TDY.
- After guard duty.
- Anytime you remove it from the arms room.

Zero your weapon with a good tight shot group. Don't be rushed at the range when you go to zero. Make sure you have a good zero. To become an expert, work on your zero shot group by:

- Taking your time at the range.
- Dry firing whenever you have your weapon.

- Going to the range as much as you can.
- Going to the range with other units.

Qualified expert should be easy after you get a good tight zero. Once you become an expert, you will have more confidence in yourself if you have to protect yourself in a war. You will get more promotion points for that next stripe that you have been thinking about.

Keep training until you become an expert and keep training after you become an expert. Once you are an expert, you may also:

- Become part of a rifle team.
- Become an inspector.
- Become a trainer.
- Become the best in the Army.

LEARN FROM OTHERS

One way to get ahead of your peers and learn how to be a good leader, before you are assigned a leadership position, is to learn from others. There are many things you can learn, such as:

- The right and wrong way of doing things.
- How to handle soldiers' problems.
- How to care for the soldiers.

The right and wrong way of doing things is very important; and it is just as important that you ensure things are done the right way and done by the book. There are times when it may take longer to do something the right way; it may seem the hard way, but it is the safe and right way. If you see another soldier doing something wrong, tell him. Rank has nothing to do with someone doing something wrong. Always let them know because:

- The only way is the safe and right way.
- You may do the same thing or teach it to others.
- It shows that you know your job and your duties.
- Someone could get hurt or killed.
- There could be damaged equipment to pay for.

There is the right way and the wrong way of doing just about everything. If you know someone is doing something wrong, it is your responsibility as a soldier to let them know it. The main thing is to stop it. You can do this by:

- Letting them know that it is wrong.
- Reporting it to your boss.
- Using the chain of command.
- Reporting it to the IG or EO.

People don't like to be told that they are doing something wrong. Be able to show them in writing what and how they are doing it wrong; know what you are talking about.

How to handle soldiers' problems is something every soldier needs to know. Learn all you can, as soon as you can because:

- You can help yourself.
- You will know what to do when your soldiers have problems.
- You will have more confidence working with other soldiers.
- Others will seek your advice or help.
- It prepares you for leadership positions.

Your goal should be to solve the problems fast and correctly. Learn from others how this is done. Some ways are:

- Make a problem solving book.
- Talk to the soldier that has the problem.
- Talk to others.

To make a problem solving book, write down all the problems you have or had, but first make a special subject for each problem. The subject must be related to the problem. Some subjects that you may have or could have are:

- No pay due
- Soldier facing punishment
- Soldier reporting to work late
- Soldier not being recommended for promotion
- Soldier didn't pass PT, SQT, CTT or other test
- Soldier is in debt

- Child or spouse abuse
- Dear John letters
- DWI, overweight, profiles, QMP and many others

Once you have the subject, write the problem that is associated with it, how it was solved and how fast it was solved. Also, you can write down other ways it could have been solved. Write down the outcome. Talk to the other soldiers that have a problem; if you can help them, do so. If not, have them see someone that you think may be able to help. There are many agencies in the Army that are designed to help solve soldiers' problems. Some of these are:

- The Army Emergency Relief
- The American Red Cross
- Equal Opportunity Committee
- Community Counseling Center
- Drug and alcohol counseling center
- Jag office
- Commissaries and Post Exchange
- Your Chain of Command and many others

Talk to others about solving soldiers' problems. You don't have to know a soldier with a problem to ask about them. Find a soldier that you feel is good at solving soldiers' problems and ask him how he would solve a particular problem. This is also a good way to get some answers in your book. Use the post phone book to find the agencies.

PRACTICE SOLDIER CARE

How do you care for the soldier? This is not hard because you care for him the same way you care for yourself or the way you would like to be cared for. Soldier care is more than a 5x8 card with the names and addresses of his loved ones or next of kin. To care for him, you have to be able to help him solve his problem. To do this, you have to know all there is to know about him. You have to know:

- What kind of attitude he has.
- How he feels about himself, family and the Army.
- What his goals are in life.

- How he feels about leadership.
- How he feels about other soldiers and officers.
- Does he have confidence in the Chain of Command?

You have to know your soldiers if you want to care for them. Because two soldiers have the same problem, it wouldn't mean you would solve them the same way. If you make a book you will write down different ways of solving them. Once you start caring and learning about your soldiers, you will know when they have a problem by the way they behave. Some signs are:

- Change in attitude
- Late coming to work
- Not associating with others
- Military bearing going downhill
- Can't get along with other soldiers

There are many more ways to tell. This is something you will learn for yourself or from talking to others. Don't wait for a soldier to come to you with his problems because:

- He may feel he can solve it.
- He may not have confidence in you.
- He may not want you to know.

A soldier has to know that you care about him before he will trust you to help him with his problems. There are ways to show you care:

- You can ask about his family.
- You can give him time off if possible.
- You can stop by to visit, on or off post.
- You can send him Birthday, Christmas or other cards.
- You can thank him for a job well done.
- You can see that he gets rewards and gets promoted.
- You can stand up for him when he needs you.

Soldier care is important to you and to the soldier; it is something you need to learn as soon as you can. There is no better reward than helping another soldier. Also, it's good leadership!

Chapter Two

HOW TO FIND OUT IF YOU ARE ELIGIBLE

In order to get promoted to the next higher grade, you will have to be eligible for promotion. The thing to do is find out if you are eligible. Army Regulation 600-200, Chapter 7 will give you most of the information you will need about promotion. Some of the things that make you eligible for promotion are:

- Time in service and grade
- The physical fitness test
- The skill qualification test
- Weapon qualification
- Height and weight
- Promotion points
- Security clearances
- Promotion recommendations
- No more than two waivers
- Leadership schooling
- Civilian education
- Alcohol and drug programs

Each soldier must have the minimum time related requirement before he can be promoted to the next higher grade.

TIME IN SERVICE AND GRADE

Private E-2

A Private E-1 will be promoted to Private E-2 when he has completed six months active duty, unless it is stopped by the commander. Time spent in a delayed entry program does not count. An E-1 can be promoted to Private E-2 in four or five months but only twenty percent of assigned and attached E-2 soldiers may have less than six months time in service. He may be promoted before time if it's in his enlistment contract. Three percent of the basic training and advance training E-1's can be promoted after they complete BCT and AIT.

Private E-3

A Private E-2 will need twelve months of time in service and four months time in grade to be promoted to the grade of E-3. Also, he can get promoted with two months in grade and twelve months in service. Six to eleven months time in service and two months time in grade can get an E-2 promoted to an E-3, but only twenty percent of the assigned and attached E-3 soldiers may have less than twelve months time in service, so a soldier can make E-3 within a year.

Specialist E-4

A Private First Class with twenty-six months time in service and six months time in grade can be promoted to Specialist. The time in grade can be waivered to three months. If a security clearance is required, it can be advanced based on interim clearances. If no accelerated advancements are authorized and none were authorized for the preceding two months, the Commander may advance one soldier to CPL/SPC without regard to the percentage restriction, but the soldier must have twelve months in service and three months in grade. The same is true for a PFC that graduates from ranger school. Commanders with zero waivers can promote any soldier that has eighteen or more months in grade, but only to Specialist that needs a time in service waiver.

Sergeant E-5 and E-6

Promotions to grade E-5 and E-6 are made against promotion point cut-off scores. Headquarters Department of the Army will determine the needs of the Army by grade and MOS. The promotion point cut-off scores for primary and secondary zone promotion to grade E-5 and E-6 are announced. This authorizes Commanders to promote the best qualified soldiers Army wide in each MOS. The promotion authority personnel may waive no more than two of the three requirements for promotion; they are:

- Time in service
- Time in grade
- Skill qualification test

Soldiers recommended for promotion in the secondary zone must be outstanding. This waiver allows soldiers who show outstanding potential through performance to be considered for promotion. To get promoted to Sergeant E-5, the soldier will need:

- 36 months time in service (primary zone)
- 18 months time in service (secondary zone)
- He can be boarded in 33 months (primary zone)
- He can be boarded in 15 months (secondary zone)
- 8 months time in grade is needed, waiver to 4 months

To get promoted to Sergeant E-6, a Sergeant E-5 will need:

- 84 months time in service (primary zone)
- 60 months time in service (secondary zone)
- He can be boarded in 81 months (primary zone)
- He can be boarded in 57 months (secondary zone)
- 10 months time in grade is needed, waiver to 5 months

The soldier can only be promoted in his primary MOS and can compete for promotion with a waiver if he scored 59 or less on his skill qualification test. Soldiers must have the appropriate security clearance or a favorable security investigation required by the MOS in which promoted. Promotion to E-5 and E-6 may be based on an interim clearance. Waiver is not granted. Soldiers must have PLDC or BNCOC to be promoted, but they can go to the promotion board before they attend the course. The soldier cannot have a record of court-martial convictions, AWOL or lost time on current term of enlistment waiver because these disqualifications may compete for promotion. A reenlistment waiver may be submitted in advance for promotion purposes. The soldiers that are in "Special Operation," with 18 months time in service and 450 promotion points, can be promoted to Sergeant E-5 the month after they are placed on the standing list, regardless of time in grade. For promotion to Sergeant E-5, Range School graduates that meet the requirement of 18 months time in service and 4 months time in grade may be promoted the first day of the third month after they are placed on the promotion standing list. Once out of Range School, the soldier will be given 999 points if he is already on the promotion standing list.

Promotion to Sergeant E-7, E-8 and Sergeant Major

A centralized promotion system has been in effect for promotion of enlisted soldiers since:

- January 1, 1969 for Sergeants Major
- March 1, 1969 for Master Sergeants
- June 1, 1970 for Sergeants First Class

Selection and promotion authority by Headquarters Department of the Army does not deprive local Commanders of the authority to reduce soldiers in grades E-7, E-8 and E-9 for inefficiency or conviction by a civil court. Eligibility for promotion consideration to grades E-7, E-8 and E-9 is based on date of rank. The Department of the Army will announce the date of rank for the primary and secondary zone. To get promoted, the soldier must:

- Meet announced requirements prescribed by HQDA.
- Have at least 6 (E-7), 8 (E-8) and 10 (E-9) years active service.
- Have 6, 8 and 10 years creditable for basic pay.

- Be serving on active duty, in an enlisted status during the board procedures.
- Have a high school diploma, GED, Associates or higher degree.
- Not be barred from reenlistment or QMP.
- Not have an approved retirement before the board convenes.
- Not be eligible for reenlistment due to a DCSS affect.
- Be a graduate of ANCOC, if E-7, with DOR of 1 April '81 or later.

The soldier in grade E-7 must have the security clearance required by the MOS he is promoted to. Before being promoted to Master Sergeant or Sergeant Major, the soldier must have a favorable NAC completed or have a secret clearance or higher.

PHYSICAL FITNESS TEST

The physical fitness test is a test all soldiers will have to pass before they can be promoted to the next higher grade. Soldiers in the grade E-7 through E-9 don't have to appear before a promotion board because they are selected by the Department of the Army, but their promotion can be stopped if they don't pass the test. The other soldiers will have to pass the test also. E-4's and E-5's will get promotion points for their physical fitness test scores (up to 50 points), but can get as much as 55 points if they score 300 on the test and get a Certificate of Achievement (5 points) with a LTC or higher signature. The soldier will have to score at least 60 in each event, for a total of 180; however an E-4 or E-5 going to the promotion board will get only 10 promotion points for a 180 physical fitness test score. Those soldiers with a permanent profile will be granted the minimum qualifying score of 60 for each waived event and use their actual score for each event taken. Soldiers that are on a temporary physical profile may use their current test score if it is not more than 1 year old. If a soldier fails the Army's Physical Fitness Test or has not taken a physical fitness test within the past nine months, he will be placed under suspension of favorable personnel action (Flag) in accordance with AR 600-31. The date of rank and effective date for soldiers that meet all promotion eligibility requirements while on suspension will be the day following the successful completion of the most recent physical fitness test unless:

- The soldier has a permanent or temporary profile that prevents him from taking the test.
- The soldier is unable to take the physical fitness test because of conditions beyond his control.

The skill qualification test is another test where the E-4 and E-5 can get promotion points. For each point the soldier gets on his skill qualification test he will get two promotion points. The E-1 through E-4 may or may not have had an SQT, but if they had an SQT it will probably have an effect on their promotion. If an E-4 or E-5 doesn't have his score in his records when it is time to appear before the promotion board, he can still go to the board if he can get it from TSO. Soldiers who have not tested or don't have an SQT score must be in one of the following categories in order to compete for promotion. They are:

- No SQT for the soldier's primary MOS.
- Testing in primary MOS has been suspended.
- Soldier was exempted by CDR, MILPERCEN.
- Soldier deferred IAW AR 350-37.
- Soldier exempted due to deferment.
- Soldier was tested but no score received.
- No previous test and current test window still open.

If it's your fault that you fail to take the skill qualification test, you will be ineligible for promotion. You will have to take the test or 12 months will have to elapse since the close of the test window. Afterward you may go to the promotion board using the no fault provisions. The skill qualification test is a very important test, not just because of the promotion points, but it shows the Commander how good you are at your job or MOS. When you go to the promotion board the Commander can give you up to 200 points. If you don't do very well on the test you could lose out on points. Anything less than 59 on the test will not get you any points and you will be unable to go to the board without a waiver.

WEAPONS QUALIFICATIONS

Weapons qualifications are not necessarily needed to get promoted from E-1 to E-4. If you are going to the promotion board, for E-5 or E-6, you will get promotion points for your weapon qualification; which are:

- 50 points for expert
- 30 points for sharpshooter
- 10 points for marksman

Now you can see how important it is to try and be an expert with your weapon. When you go before the promotion board, your most recent assigned weapon qualification will be used, unless you fail to qualify through your own fault. If you fail to qualify, you will not get points during the time you are boarded. Once you qualify, your points will be added to your promotion points during the next recomputation, unless you have more than 35 points which can be added then. The entry in item 9 on your DA Form 2-1 may also be used to award points. You should always make sure this is updated. Also, it is important to know that if you can qualify expert before you go to the promotion board and qualify sharpshooter afterward, you will lose 20 points during the next recomputation, unless you fire expert again before the recomputation. Firing expert should always be your goal.

HEIGHT AND WEIGHT

Being overweight will keep you from getting promoted if you exceed the body fat standard or maximum allowable weight established in Army Regulation 600-9 and no underlying or associated disease has been found to cause the overweight condition. Promotable status will be regained when the promotion authority determines that you are no longer required to be in the overweight program. Get a copy of AR 600-9 and find out how much you can weigh for your height and sex. You should also find out how much body fat you are allowed. That way you will know if you go over a pound for your weight. KEEP IT DOWN!

PROMOTION POINTS

Promotion points are needed if you are to be placed on the promotion board standing list. E-4's have to have at least 450 points and E-5's have to have at least 550 points. These are the amounts of points you will need before you can be placed on the promotion standing list. Your Commander can give you up to 200 points before the board; you can get up to 800 administrative points and 200 points off the board, for a total of 1000 points.

SECURITY CLEARANCES

Security clearances are needed for promotion from E-4 through E-9. Some E-2's and E-3's may also need a clearance. Soldiers on the list for promotion to the grades E-8 and E-9 who do not have a favorable National Agency Check (NAC) completed or a final secret security clearance or higher will not be promoted. An E-7 must have the security clearance required by the MOS (job) in which he will be promoted. Since Headquarters Department of the Army administers promotions to grades E-7 through E-9, Commanders are responsible for notifying HDQA when soldiers whose names appear on a DA recommended list are non-promotable. Soldiers in the grades E-1 through E-6 must have the appropriate security clearance or favorable security investigation required by the MOS in which promoted. They may be promoted based on an appropriate interim clearance. Keep your clearance up-dated!

PROMOTION RECOMMENDATIONS

Promotion recommendations are needed for promotion to the grades E-2 through E-6. Recommendation for promotion, including waivers, will be prepared on DA Form 3355 by using the remarks section for E-4's and E-5's going before the board. Recommendation for E-2 through E-4 may be:

- Done on a counseling statement
- Done orally
- Done by voting
- Done by Commander

Most of the time, for promotions to the grades of E-2 through E-4 the First Sergeant and the Platoon Sergeant will get together and select who will be promoted. The First Sergeant will take the list to the Commander for his approval or disapproval. All recommendations must be processed through the Unit Commander or the Soldier Parent Organization. The promotion authority for E-5 and E-6 must be in the grade of a Lieutenant Colonel or higher; he will approve or disapprove the recommendation for promotion. E-7 through E-9 do not need a recommendation for promotion because they are promoted by HQDA. If they are placed on the list for promotion and for some reason should not be promoted, it is up to their Commander to send in the paperwork needed to HQDA to stop the promotion.

WAIVERS

No more than two waivers may be used for promotion. These include:

- Time in service
- Time in grade
- Skill qualification test

For time in service, an E-4 going to the promotion board for E-5 can get a 15 month waiver; an E-5 going to the board for E-6 can get a 24 month waiver. A waiver for time in grade for an E-4 going to the board can be four months; an E-5 going to the board for E-6 can get a waiver for 5 months. The waiver for SQTs is the same:

- He may not complete a waiver if he scored less than 59.
- He has no SQT for his primary MOS.
- He was exempted by Commander of MILPERCEN.
- He was deferred and sixty days have elapsed.
- He was exempted due to deferment.
- He was tested but no score has been received.
- He has no previous test and current window still open.

LEADERSHIP SCHOOLING

Leadership schooling for promotion to grades E-5 through E-9 is very important and should be done as soon as possible. Never let anyone stop you from going; use the Chain of Command if you have to, or whatever it takes, but go to school. It is important to you and your career. There are many times when a good soldier is held back from school because someone tells him they can't let him go at that time, or that he is too important to let go at that time. If someone should tell you this tell them to pretend that you are dead. Just think, if you go on leave they will do without you. They can do without you while you go to school and you can move up in rank as they did. The same people who tell you that they need you will stop at nothing to go to school themselves if they could. As soon as you make E-4, try to go to a leadership school (PLDC), then BNCOC, ANCOC and the First Sergeant and Sergeant Major Academy. You will need these schools before you can be promoted. If you are going to the board for E-5, you can go before you go to PLDC. The Department of the Army picks the soldier for school in grades E-6 through E-9, but you can always write or call them and let them know that you want to go. Keep your records up to date and write or call them; they can't say anything but yes or no. You will not be able to make E-8 until you go to ANCOC and you can go as an E-6; the sooner the better. Don't wait. If you have to go to a promotion board for promotion, you can get up to 150 points for military education. Don't just wait for a leadership school, ask to go. They are good for points and they will look good in your records when it's time for you to make E-7 through E-9. Some board members at DA feel that the more schooling you have the better Platoon Sergeant, First Sergeant or Sergeant Major you will make. Go to school whenever you can and as often as you can.

CIVILIAN EDUCATION

Civilian education can help get you promoted if you are going to the board for E-5 or E-6. You can get up to 100 promotion points just for civilian education. If you had a high school diploma or GED when you entered the Army you will not get points for it, but if you get your high school diploma or GED after you enlist in the Army, you will get 10 promotion points for education improvement. Also, if you improve your GT score you get 10 points. You can get points for:

- Business schooling
- Trade schooling
- College
- Correspondence Courses
- CLEP examination

Get as much education as you can while you are in the Army. Go to your education center and see what they have to offer.

ALCOHOL AND DRUG PREVENTION AND CONTROL PROGRAM

Alcohol and drug prevention and control program will stop you from getting promoted because you will be flagged as soon as you go on the program, if not sooner. However, if you are eligible for promotion while you are on the program, you will be promoted as soon as you complete the course and your date of rank will be the same as your peers that were promoted when you would have been promoted. Army Regulation 600-200, Chapter 7 is the regulation you want to check about your promotion eligibility. You will also need to check with your boss, the First Sergeant and the Promotion Clerk who more or less keep up-to-date about promotion. The more you know about promotion, the better prepared you will be and the sooner you will make your rank. Remember, it's nothing to brag about how much time in grade you have. Brag about how soon you made Command Sergeant Major!

Chapter Three

HOW TO GET PROMOTION POINTS

If you are in the grades of E-1 through E-5, this may be your favorite chapter. Here, you will learn how you can get promotion points and how many you can get from the many areas.

Before you can get promoted to the grades of E-5 and E-6, you will have to be recommended by your Commander and a majority of the voting members of the promotion board. You still may not be promoted if you don't have the points needed to be placed on the promotion standing list, which is 450 points for the E-5 and 550 points for the E-6. On the other hand, you can have 450-1000 points and still not be promoted if you are not recommended for promotion. The Commander is the first person whose recommendation counts. The Platoon Sergeant's, Platoon Leader's, or other's recommendations let the Commander know that someone thinks the soldier should be promoted. If the Commander feels the same way, he will recommend the soldier. With the correct amount of points and the board recommendation, the soldier will be placed on the promotion standing list. There are many reasons that a Commander may not recommend a soldier for promotion. Some are:

How to Get Promotion Points 47

- He won't give the soldier a waiver for time in service.
- He won't give the soldier a waiver for time in grade.
- He won't give the soldier a waiver for SQT.
- The soldier has no leadership schooling.
- The soldier didn't qualify with his weapon.
- The soldier didn't pass his physical fitness test.
- He doesn't feel the soldier is ready.

Any of these reasons can keep you from being recommended by the Commander and there can also be other reasons, it's all up to him. He can just say no and not sign the DA Form 3355. Some of the reasons a voting board member may not recommend you are:

- You have a sloppy appearance.
- You have no self confidence and speak too low.
- You have a bad attitude for a leader.
- You do not know enough about basic soldiering.
- You do not know enough about military programs.
- You are not informed on world affairs.
- You have poor military bearing.
- Your conversational skills need improving.

If you are recommended by your Commander to go before the promotion board, chances are very good that you will be recommended by the board members. Board members want you to be promoted just as much as you do, but they will not give it to you. All you have to do is show them that you will make a good leader. That's what it's all about. Now that you know that you can get the recommendation, let's see how many points are out there waiting to be picked up by you. You can get up to 800 administrative points and 200 points from the promotion board. The areas and amount in each area are:

- Skill Qualification Test   200 points
- Duty Performance   200 points
- Military Education   150 points
- Civilian Education   100 points
- Military Training   100 points
- Awards and Decoration   50 points
- Promotion Board   200 points

There are still more administrative points you can get before you go to the board, unless you have 800 points. We will now see how the points are broken down in each area.

SKILL QUALIFICATION TEST

Here you are looking at 200 promotion points, and if you score 100 on your SQT you will get the 200 points. The chart below shows how many points you will get for SQT if your last SQT is older than 24 months from the date scored, or if you do not have a score at the time of reclassification or recomputation. To find out how many points you will get, you will add your:

- Duty performance points
- Awards and decoration points
- Military education points
- Civilian education points
- Military training points

After you get the total from the five areas, look at the chart. Find your score and the number to the right of your score is the amount of points you will get for promotion:

Score		Points
0-37	=	120
38-75	=	125
76-112	=	130
113-150	=	135
151-187	=	140
188-225	=	145
226-262	=	150
263-300	=	155
301-337	=	160
338-375	=	165
376-412	=	170
413-450	=	175
451-487	=	180
488-525	=	185
526-562	=	190
563-599	=	195
600	=	200

How to Get Promotion Points 49

Only the points you get before going to the promotion board will be in this area. Do not get it confused with the board points.

DUTY PERFORMANCE

You can get another 200 points for duty performance (Commander points). The Commander awards you points by considering your leadership and personal traits, such as:

- Your potential for advancement
- Your attitude
- Your responsibility
- Your initiative
- How well you adapt to changes
- Your expression
- You as a whole

How many points he gives you for each area is up to him. Some Commanders:

- Give 200 points to all soldiers he recommends.
- Talk to the Section Chief or Platoon Sergeant.
- Consider your SQT, weapon and PT scores.
- Interview the soldier before deciding.
- Evaluate your leadership potential.
- Talk with the Platoon Leader.

There are no set rules for him to decide, but if you want to know, ask him.

MILITARY EDUCATION

The 150 promotion points you get in this area are not hard to get. It may take longer to get some of the points because you will have to wait for your schooling. The best way to start getting promotion points for military education is to take a correspondence course. It's best to take one on your MOS because:

- You already know something about the subject.
- It will help you on your job.
- It will help you with your SQT.
- It won't take long to complete.
- It's a fast way to get points.

You can get one promotion point for each five credit hours, and that's every five credit hours. It you have 13 credit hours, you will only get 2 points. Another way to get points for correspondence courses is to study with a group. Find some other soldiers who are trying to get points for promotion. Each of you take a course on your MOS. They should not be the same course. After you complete the course, team-up and send-off for the same course that your partner took and he will send for the course you took. When they come back, each of you get together and help each other with the course. The more you continue doing this with other soldiers in your unit, the more points you will get and the more you will know about some of the other jobs in your unit. Don't forget, (5 credit hours = 1 promotion point). Next, you get 30 points for a leadership school. Some are:

E-5	E-6	
PLDC	PLDC	Special forces course
PNCOC/CA	BNCOC/CA	Ranger School
PLC	PLC	
PTC	BTC	

For all other courses successfully completed and of at least one week duration, you will get 2 points per week, such as head start. If you are going to the board for E-6 and had PLDC when you were an E-4 or E-5, or you have BNCOC, you would get 30 points for the leadership school you attended. If you took a leadership course by mail, you will get 3 points for each week of schooling. In other words, for a four week school you would get 12 points, not the 30 points you would have gotten if you had gone to the school. There are many military schools that you may be able to attend. Talk to the training NCO and your boss about going to one.

CIVILIAN EDUCATION

You can pick up another 100 points for civilian education. did you know that all you have to do is improve your GT score? You will not only get the new score, but 10 promotion points. If you get your high school diploma or GED while on active duty, you will get 10 promotion points. You can also get points for:

- Accredited Colleges or Universities
- Correspondence Courses
- Extension or Residences Courses
- CLEP General Examination/SST
- Trade Schools
- Business Schools
- Commercial Correspondence

To find out how many points you will get for a particular school, go by your Education Center and talk to one of the counselors. They will be a big help. You need to do this before you do any type of schooling because you can also get points for your MOS. If you work in personnel, you will get points for management. All MOS's get points for physical fitness and you don't have to make 300 on your PT test to get them. Never say you don't have time to go to school because of your job. If you have a VCR there are some courses on tape you can take. If you can get 12 or more soldiers in your unit to take a class, they will come there and teach it. I knew a Sergeant that had classes in the mess hall for his cooks; they all got college educations and promotion points. Talk to the counselor and see what he can do for you. Everyone should check to see what their reading level is because that is going to be a big thing in the Army. Soon you will have to read at a tenth grade level to attend some leadership schools.

MILITARY TRAINING

All you have to do is qualify expert with your weapon and max the physical fitness test and you will get the 100 points for military training.

Weapon Qualification		Promotion Points
Expert	=	50
Sharpshooter	=	30
Marksman	=	10

Physical Fitness Test		Promotion Points
300	=	50
299-290	=	45
289-280	=	40
279-270	=	35
269-260	=	30
259-240	=	25
239-220	=	20
219-200	=	15
199-180	=	10
179-000	=	0

AWARDS AND DECORATIONS

Here you can only get 50 points; but again they are not hard to get. I will only list the awards that most soldiers will get. They are:

- Meritorious Service Medal = 25 points
- Army Commendation Medal = 20 points
- Army Achievement Medal = 15 points
- Good Conduct Medal = 10 points
- Driver Badge/Mechanic = 5 points
- Air Assault Badge = 5 points
- Certificate of Achievement = 5 points

PROMOTION BOARD

There are many soldiers that feel that the 200 points you get from the promotion board are the hardest points to get. The main reason for this is that many soldiers do not know what to expect when they go before the board. Each voting member on the promotion board will have a DA Form 3356 he will use to evaluate and recommend the soldier going before the board. To recommend the soldier all the voting member has to do is put a check mark in the block "I do" or "do not" recommend the soldier for promotion. Most of the time, this block will be filled in after the voting member has asked you his question. There are times when a voting member will fill it in before he asks a question. The soldier appearing before the promotion board will be evaluated in six categories, which are:

- Personal appearance, bearing and self-confidence
- Oral expression and conversational skills
- Knowledge of world affairs
- Awareness of military programs
- Knowledge of basic soldiering
- Soldier's attitude

Personal appearance, bearing and self-confidence will be checked beginning the minute you march into the board room. Each member will be watching to see if you march from the door to the point where you report to the president of the board, or if you walk. Make all your movements military; march! After you report to the president of the board he may give you some facing movements. They may be:

- Right face
- Left face
- About face

As you execute the movements the voting members will check out your appearance and bearing. They can also start checking out your self-confidence by the way you make the movements. Do you have to think about what you are doing or do you seem to be afraid? The president may also march you about for three to six steps (whatever it takes to get there) into a position where the voting members can check you out. His commands may be:

- Four steps forward, march
- Rear, march
- Right step, march
- Left step, march

Be sure all your movements are military; that's very important and can help you get 30 points. The way you move, stand and sit is all part of your bearing. Swing your arms when you are marching; move about like a toy soldier. Your goal is to walk out with a Recommendation and 200 points. Oral expression and conversational skills begin when the president of the board asks you to tell the board members something about yourself. This is the most important part of the promotion board because:

- You get a chance to sell yourself.
- You can get up to 75 points.
- You will get a feel of how things are going.

Selling yourself is so important that, if you are not good at it, everything can start going downhill from that point. This is because:

- You may lose confidence.
- Board members may change their attitudes.
- You could lose military bearing.
- You could get kicked off the board.

Please read "How to Sell Yourself" in the Appendix at the end of this book. Read it as many times as it takes for you to know how to sell yourself by heart.

Now, let's keep going. You can get up to 75 points when you start telling the board members something about yourself because, if you are good at it, the voting members may give you 35 points for conversational skills and 40 points for soldier's attitude. That's why it is so important to know how to sell yourself. You will get a feel for how things are going when you start telling about yourself. Oral expression is how loud you speak, how you say what you mean and how you go from one subject to another. We will talk more about this in "How to Sell Yourself." Knowledge of world affairs is, more or less, a free 25 points but it can also make or break you. Why is this so important? Well, a good leader must be informed about what is going on in the world. He has to be able to keep his soldiers informed about what is going on in the world. To do this:

- Watch CNN on TV.
- Read your post paper.

- Listen to the news on the radio.
- Read a good newspaper each day.
- Talk to others about what is going on.

Sometimes the president of the board may ask you what is going on in the world. If he does, the voting members will grade you at that time. One of the voting members may ask you about a certain thing happening in the news and, most of the time, the question asked will have something to do with the military. If you are asked to tell what is going on in the news, tell about the main event at that time, something about sports, something about the weather, or a thing or two about business or post news. You will need to be aware of military programs. If you are going to the board for E-5 or E-6, the more programs you are aware of, the more you can tell your soldiers about. There is a very long list of military programs for the soldiers. Be sure to impress the voting members by letting them know about the ones you are familiar with. To find out about them:

- Ask the First Sergeant.
- Look in the post phone book.
- Ask other soldiers.
- Check with ACS.
- Ask your chaplain.

Some programs that most soldiers and voting members are familiar with are:

- Army Community Service
- Army Emergency Relief
- American Red Cross
- Advanced Skills Education Program
- Basic Skills Education Program
- CLEP (School)

Be sure that you are aware of what these programs do for the soldier. Soldiers going up for E-6 should know about more programs. Military programs are good for 25 points and are not hard. Knowledge of basic soldiering is just what it says, basic. The board members won't try to ask anything that will make you lose points. On the other hand, you should know the basics about everything you teach your soldiers and that is what these questions are all about. The thing to remember here is that the voting member who asks you a question on basic soldiering will grade you for the answer. You will also have to know about:

- Land navigation, survival and night operation
- Inclement weather, adverse environment and terrain

Soldier's attitude can get you a big 40 promotion points. As I told you before, if you can do well in selling yourself you can pick up these 40 points when asked to tell something about yourself. No one asks you questions about soldier attitude. What the voting members do is look into your records. They are looking for:

- Letter of recommendation from your boss
- Any kind of awards
- The kind of leader you are
- Trends in your performance
- Anything about your leadership
- Something about your potential

If you can get a letter of recommendation from your boss, do so. I'm not talking about the Commander because he recommends you on the DA Form 3355. If you do work for the Commander, ask him to write you a letter of recommendation. The letter should tell about your:

- Duty performance
- Awards you received in the unit
- Your potential for advancement
- Something important you did for your soldiers

Everyone going to the board should try to get a letter from his boss or another soldier in his chain of command. It is not a must that you have a letter, but if you do, most of the time the voting members will read the letter instead of looking through your records. Each voting member on the board can give you up to 200 points. Once all the points are totaled, they will be divided by the number of voting members. That will give you your score from the board. Get 200!

Chapter Four

HOW TO PREPARE FOR THE PROMOTION BOARD

In order to be promoted to E-5 and E-6 a soldier has to appear before a promotion board. The boards are conducted before the 15th of each month if there are soldiers recommended.

The board could be composed of Commissioned Warrant Officers and Non-Commissioned Officers, but in most cases you will find the board will be made up of all Non-commissioned Officers, with a Command Sergeant Major serving as the President of the Board. A Sergeant Major can be President if there is not a Command Sergeant Major. Most likely, the members would be the First Sergeants from that Battalion or Brigade. There can be other NCO's serving as board members as long as they are senior in grade to those being considered for promotion. There will also be a recorder from the personnel section or someone well qualified in military personnel procedures. However, the recorder does not vote so he can be of any grade. There should be at least one voting member of the opposite sex of those being considered. If not, this should be recorded in the board proceeding by the recorder. Minority ethnic groups will be appointed as voting members, but by no specific number or ratio as long as they are not all from a minority ethnic group.

There should be at least three voting members and a recorder. The President of the Board can choose to be a voting member or to vote only to break a tie, but he can only vote once. Each voting member will complete a DA Form 3356 (Board Member Appraisal Worksheet) on each soldier being considered for promotion. He will mark the proper block to recommend or not recommend each soldier that appears before the board. So you see, a soldier can answer all the questions and still not be recommended for promotion.

After each soldier appears before the board, the recorder will collect the DA Forms 3356 and let the president know the results of the votes. If a tie exists the President of the Board will vote to break the tie. The Board President will sign all DA Forms 3355 (Promotion Point Worksheet).

The board must be comprised in such a way as to avoid a possible tie in voting. If the president votes on all the soldiers, the total voting members (including the president) must be an odd number. Once the board convenes the members cannot change.

Soldiers competing for E-6 need at least 550 points after the board to be placed on the promotion standing list. Soldiers competing for E-5 need 450 points to be placed on the promotion standing list. Soldiers not recommended for promotion will be counseled about areas that need improvement.

Now that you have an idea how a promotion board operates, what can you do to prepare yourself for the board? First of all, you will need to:

- Learn all you can about the board members.
- Make a study guide.
- Check your records.
- Study your promotion guide.
- Prepare your uniform.
- Request a pre-board.

LEARN ABOUT BOARD MEMBERS

Learn all you can about the board members because this can help you face them at the board. It will help if you know:

- Their name and sex.
- If they are permanent board members.

How to Prepare For the Promotion Board

- Who their replacements are.
- What subject(s) they will have.
- Where they are seated on the board.
- If they try to be hard.
- What unit they are from.

You will want to know the names of the board members so you can be sure you can pronounce them correctly. Believe me, you can lose a point or two if you pronounce them incorrectly. How do you feel when someone pronounces your name wrong? If you know you will have a board member whose name you can't pronounce, you will have time to find out the right pronunciation.

If you go to the board and find out there is a member you didn't know about, find out who he is. You can ask:

- The member before the board.
- The recorder of the board.
- The President of the Board.
- Someone from his unit; call if you have time.

If you find yourself sitting in front of a board member and he asks you a question and you can't pronounce his name, don't try; just say, *"Excuse me, First Sergeant (or whatever his rank may be), but can you tell me how to pronounce your name?"* If you get it wrong after he tells you, he probably won't take points from you because you pronounced it wrong.

You will want to know what sex the board members are because, for some reason, there are board members who are a little harder on the opposite sex. Also, if there is a female on the board, she may be a permanent member of the board. Find the same things out about her that you would about the male members.

It's important to know if they are permanent members because:

- They will be more relaxed.
- They know their duties.
- They can help you.
- They sometimes use the same questions.

Permanent board members will, from time to time, be replaced for one reason or another. Some of the reasons are:

- They go on leave or TDY.
- They have other commitments.
- They PCS before the board.

The thing to do here is find out all you can about the replacement member. If the same person replaces a particular permanent member all the time, that's not so bad because:

- He knows something about the board procedures.
- He may have the same subjects and questions.
- He knows the other board members.
- He is more relaxed on the board.

If the replacement is not a "permanent replacement" or doesn't normally replace the permanent member he may:

- Not know much about board procedures.
- Not know the basics about the subject.
- Have found out about the board at the last minute.
- Try to be hard to impress the board President.
- Not be relaxed like the other board members.

What subject will be presented at the board is important. You can find this out in the memorandum you will be given before you go to the board. It will tell you who the board members are and what subjects they have. But why wait? If you know the board members are permanent members, get some old board memorandums. You can:

- Ask a board member for them.
- Ask the recorder.
- Ask soldiers that have been to the board.
- Check with your personnel section for copies.
- Ask a soldier that you know is going to the board.

Try and get to the board as it is being set up. Most of the time this is done the night before or early on the morning of the board. You will know where the board will be held because it will also be on your memorandum. There will be name plates on the table where the board members will sit. The board president can call on a member to ask questions in any order he pleases, but you will want to know where they are sitting so you will know:

- Where the replacement is sitting (if there is one).
- Where any hard core member is sitting.
- Where the president is sitting (center most of the time).

If you know of a member that tries to be hard, find out what he is hard about. It may be:

- The uniform
- The male or female soldier
- Map reading or his other subjects
- Self confidence
- Soldier's attitude, etc

If you find out what he is hard about—good, be prepared. The next thing you may want to know is what unit the members are from. This will also be on the memo when you get one, but it's good to know ahead of time because:

- You can ask someone in his unit about him.
- His replacement will probably come from his unit.

MAKE A STUDY GUIDE

Make a study guide to help you prepare for the promotion board. You will make it similar to the guide made for guard duty. You will need the names of the president, the other board members and the replacement(s), if any. Also, find out if there is a female member and, if so, who she is. Is she a permanent member? If so, who is her replacement?

Once you get all the names and subjects, all you need to do is try and find out what questions they ask. To do this:

- Ask the members after a board.
- Ask the soldiers that have been to the board.
- Pay the recorder to write them down for you.
- Ask a soldier after every board.

For your study guide, you will need to write down other areas you will be graded in. Some are:

- Personal appearance, bearing and self-confidence.
- Oral expression and conversational skills.
- Knowledge of world affairs.
- Awareness of military programs.
- Land navigation, survival and night operations.
- Inclement weather operation, adverse environment and terrain.
- Soldier's attitude.
- Leadership and potential for advancement.
- Trends in performance and chain of command.

Each member will have a DA Form 3356 and will grade you in all areas, but the member that asks you questions about his subjects will grade you on "Knowledge of basic soldiering."

Each month you should have something to write in your study guide about promotions. Keep your guide up-to-date.

Your guide can also be used for Soldier of the Quarter and other boards because:

- Most of the boards have the same members.
- Some have the same subjects and questions.
- They are sometimes done at the same time as promotion.
- Members may not take the time to prepare new questions.

CHECK YOUR RECORD

Check your record before you are told to do so. A day or two before you go to the board, you will be given time to check your record. You should check your record before then because then you will have more time. When personnel makes an appointment for you to check them, you will find that:

- Other soldiers are doing the same thing.
- You may be rushed.
- The clerk may have other records to work with.
- Changes may have to be made in your record.

If you check your record a month ahead of time, you can be assured there is plenty of time for additions or corrections. You may check or add:

- GT score
- Awards
- Civilian and military schooling
- Height and weight information
- SQT scores

STUDY YOUR PROMOTION GUIDE

Now that you have made a study guide, and you have the best one, you know what subjects will be covered and some of the questions you may be asked. Don't rely on just the questions you have in the guide. Learn all you can about each subject. That's still not much if you look at some of the guides the other soldiers have. You have the best study guide because you know what subjects will be covered and, maybe, who will ask questions about them. Once you have your guide put together, work on it each month and study, study, study. You can study:

- When you have C.Q. or other duties.
- When you are on leave.
- During your lunch hour.
- On the weekends.
- When you are on TDY.
- When you are at work or an hour before work time.

The more you study, the better you will do and the more confidence you will have in yourself. Study your guide every chance you get.

PREPARE YOUR UNIFORM

Prepare your uniform as soon as you find out you are going to the promotion board. This should not be hard because your Class A Uniform should be prepared at all times.

The first thing you will want to do is have the uniform cleaned. Even if it is already clean, have it cleaned again.

Have your uniform checked out before you have it cleaned. It will look good to you so let others check it for:

- The way it fits.
- The way the trousers are cut and how they hang.
- How long the skirt is above or below the knee.
- How the stripes are sewn on.
- How long the sleeves are.
- How the brass, medals, awards and name plate look.

Get AR670-1 and set the uniform up according to the regulations and make an appointment to have it checked by three soldiers:

- The Section Chief
- The Platoon Sergeant
- The First Sergeant

If one asks why you went to see someone besides them, tell them that you wanted to see if they knew as much about the uniform as they know. Do not wear gold-plated brass to the promotion board, or shoes that you just wiped off to make them shine; that's the lazy soldier's way. If you can't shine shoes or boots as well as you would like, read "How to Spit Shine Boots" in the Appendix of this book.

REQUEST A PRE-BOARD

Requesting a pre-board is the next best thing to going to the board. There are other boards you can use as a pre-board. Some are:

- The Soldier of the Month Board
- The Soldier of the Quarter Board
- The Soldier of the Year Board
- One-On-One Boards with another soldier

Some units may not have pre-boards, but most will form one if you ask. All you need is someone to act as the president and three other soldiers. Try to get it set up just like it will be at the real board, or have it at the same place. See if you can get the board recorder to come to the pre-board. Pay him if you have to because he can tell you how things will be at the real board. To get your pre-board, talk to:

- Your Section Chief
- Your Platoon Sergeant
- Your First Sergeant
- Your Sergeant Major
- Other NCO's in your unit
- First Sergeants in other units of your Battalion

If the unit doesn't have pre-boards, talk to them about it. Tell them:

- It will help the soldier get promoted.
- It will help the soldier during other boards.
- It will help during command inspections.
- It can train NCO's about board procedures.
- It makes the chain of command look good.

The pre-boards are good training for you and other soldiers because:

- They are harder, but you get good training.
- You will get used to facing your superiors.
- You will have more confidence in yourself.

Also, soldiers that have been to a board can be part of a pre-board.

Chapter Five

HOW TO CONDUCT YOURSELF DURING BOARD PROCEDURES

Now that you have arrived for your day at the board, what will it take to walk out with 200 points and a recommendation for promotion? All it takes is for you to sell yourself to the promotion board members. To do this you must:

- Dress in accordance with AR 670-1
- Have knowledge about the board members
- Have knowledge about basic soldiering
- Be able to express yourself well
- Show respect for the board members rank
- Depart correctly

DRESS IN ACCORDANCE WITH AR 670-1

By now your Class A Uniform should have been cleaned for the promotion board, even if it was hanging up clean. Also, have the cap cleaned and always buy the latest style Class A Uniform and Cap.

Carry your uniform to the board. Don't wear it, even if you are the first soldier to appear before the board members. Once you put it on,

How to Conduct Yourself During Board Procedures

don't sit down until you are told to do so by the President of the Board.

If you are a male soldier, make sure your trousers have the West Point Cut. If you are a female soldier, find out if you are to wear the skirt or the pants.

Since the uniform is like a background for your awards, medals, ribbons and other accessories, make sure it's cleaned and pressed.

Also, you will need:

- A long sleeve 415 green shade shirt
- A spit shined pair of low quarters
- A four-in-hand necktie
- A new pair of socks and underwear

You should wear the long sleeve shirt with the Class A uniform. Again, have it cleaned and pressed.

Before you put on your coat, be sure you are wearing a name plate, shoulder boards or pin on rank.

Spit shine your low quarters; don't wear the ones that are self-shining. If you are not sure how to spit shine you shoes, read "How To Spit Shine Boots" in the Appendix.

This is also true for gold plated brass; don't wear it to the board. The members of the board want to see what you can do for yourself, not what you can have done for you.

Since you only get one shot at the board, unless you go back for more points, look your best.

Wear the long four-in-hand black tie. Yes, have it cleaned. If you can't tie it, have someone tie it for you. Don't wear the clip on tie; the members can tell.

You will want to buy new socks and underwear; it will make you feel cleaner and the socks are lint-free. Don't try to fool anyone by wearing the brown T-shirt. Sometimes the president may ask to see it and the brass on your belt.

Remember, when you go inside the board room, leave your cap, gloves and handbags outside.

Female soldiers should also have their uniforms checked out. Try to get one of the board members to check it for you. Believe it or not, not many male soldiers know much about the female uniform, so if you can get one of the voting members to check it, that may help.

Once a soldier leaves the board, the members always ask each other about something they are not sure of. If it is wrong, they may say that they told you that was the right way and you won't lose points. You make sure you know how it should be because they may ask you how it should be. Remember AR 670-1!

BECOME KNOWLEDGEABLE ABOUT BOARD MEMBERS

The more you know about the board members, the closer you will feel toward them and the better you will perform on the board. We all fear the unknown, so get to know them or about them and you won't be nervous when time comes for them to question you. You don't have to know much about them, just things like what unit they are from and if they are permanent board members. If you know anyone in their units, relate them to that person and you will feel more relaxed about them as a voting member.

KNOW BASIC SOLDIERING

If you know the subjects well, you will have confidence in yourself when you face the board members. If not, you may feel tense when asked questions about subjects you are not sure of.

If you made a study guide and learned all you could about the subjects, you should do well when asked questions about them. Be prepared!

EXPRESS YOURSELF WELL

After you report to the President of the Board and your appearance has been checked, he will ask you to have a seat. Then he will ask you to tell the board members something about yourself. Once you speak the first word, speak as if you are talking to two ranks of soldiers behind the board members. This way, you can be sure they will hear you.

If you don't know the answer to a question, don't start speaking low. Just let the board member know you do not know the answer to the question. When you speak low it gives the impressions that you have no confidence in yourself. So speak up. Don't forget—loud enough to be heard two ranks behind the voting members.

SHOW RESPECT FOR THE BOARD MEMBERS' RANK

Most promotion boards will have all enlisted members, with a Command Sergeant Major as the President of the Board. There are only three titles for Sergeants. They are:

- Sergeant Major
- First Sergeant
- Sergeant

When you report to the President of the Board, don't say Command Sergeant Major. There are some Sergeants Major that feel big when you call them Command Sergeant Major. So find out what he likes. Ask some of the voting members or some of the soldiers that went to the board before you.

The President of the Board will introduce you to the board members, so whatever rank he calls them, you do the same. Sometimes there may be an E-7 First Sergeant, if so, call him First Sergeant if that's what the President calls him. You can also look at the name plate in front of the member to see how you should address him.

DEPART CORRECTLY

After all the voting members have asked you their questions, the board president may ask you more questions. Most of the time, this is just to clear up something that he may think one of the members may not be sure of. He may then ask the voting members if they have any more questions for you. If you answer a question wrong, the member may ask you again. He may also ask the question a different way. They will only do this to help you get more points.

The Sergeant Major may also ask you if you have any questions for the members. If so, ask them.

If there was a question you were not sure of, you can ask them to repeat the question but this should be done when the question is first asked.

If you know you missed the answer to a question and you know the right answer now, ask them to repeat the question. You may get more points for it the second time, whether you answer right or wrong.

If you have questions for the board members, ask them. If not, thank them for taking the time to prepare the board for you and the other soldiers. This can also get you points, especially if you are almost the last soldier and the others didn't thank them.

You don't have to thank them; it's just a way to get more points and show them you care. Thank them!

After you thank them, the president will tell you that's all. As soon as he lets you know it's over, jump to attention as if a Colonel just walked into the room, march back up to the desk, salute and wait for him to return it. Now sharply bring down the hand to your side and do:

- Right face or half right
- Left face or half left
- About face

or whatever facing movement will point you toward the door, march to the door, open it and go out.

I've seen soldiers go from 200 points down to 195 points or less just because they didn't:

- Salute the President before leaving
- March, but walked up to his desk
- Stand at attention when they got up from the chair
- Stand at attention in front of his desk
- Make facing movements before departing
- Wait for the president to return his salute
- March, but walked out of the door

It's not over until you are back into BDU's. So don't give away points when you are leaving. I have seen soldiers almost run out. Some soldiers call the chair the hot seat.

All you have to do is have it all together and you will get the 200 points and recommendations for promotion. Be sure to read "How To Sell Yourself" in the Appendix at the end of this book and read the book again.

Chapter Six

HOW TO PREPARE FOR E-7, E-8 AND E-9 PROMOTIONS

The Department of the Army will send a message to the Military Personnel Office announcing the zone of consideration. The Military Personnel Office will send a letter to the Battalion Personnel Section asking them to send the soldiers in the zone of consideration to the Military Personnel Office to review their records. To get ahead of your peers for promotion you will need to:

- Check your records
- Double check your NCOER
- Make a photo
- Send a letter to the board president

CHECK YOUR RECORDS

Once you are told to report to the Military Personnel Office to review your records, be sure to take along the latest copy of your NCOER, physical and proof of any schooling or awards you may have.

When you check your Form 2A, check everything. Pay close attention to:

- Your name and social security number
- Date of rank and PMOS/ASI
- PULHES and physical cat code
- GT score, military and civilian education codes
- PEBD/BASD and ETS
- Date of last photo and NCOER verification
- Date of birth
- Ethnic and citizenship code
- Spouse social security number
- SMOS/ASI and duty MOS/ASI
- SQT date, score, MOS, rating and percentage
- Date eligible for good conduct award

Take time and check the entire record. If you don't understand something, ask someone. When you check Form 2-1, check these items:

- 1 and 2   Name and social security number
- 5   Overseas service
- 8   Aptitude area scores
- 9   Awards, decorations and campaigns
- 17   Civilian education and military schooling
- 18   Appointments and reductions
- 19   Specialized training
- 22   Physical status, "up-date it"
- 23 and 25   Personal and family data
- 35   Current and previous assignments, "up-date it"

Again, check the entire record and up-date anything that needs to be up-dated. After you are sure of everything, sign the record. If you don't, it will be sent to the Department of the Army and the promotion board. If you don't take the time to check and sign them, why should they take the time to promote you? If you can't take care of yourself, how can you take care of your soldiers? This is the way some board members think.

DOUBLE CHECK YOUR NCOER

DA Form 2166-7 is your NCO Evaluation Report (NCOER). It is the most important part of your records for promotion and assignments. If you don't have a blank one, get one and look at it; look it over very carefully.

Part one is the administrative data; here you want to make sure your name and social security number are correct. Another soldier may have the same name as you but will not have the same social security number. Sometimes soldiers' records are mixed up because of the same name, so make sure yours is correct. It is up to you to keep your records up-dated. You should send for your micro-fiche no less than once a year or three months after you get an NCOER, promotion, schooling or an award to see if it is on your records. It won't do you any good to have these things if they are not on your micro-fiche. Go by DA and check your records. Take anything with you that you want added on; they are very nice people.

Part two is authentication. You should know your rater, senior rater and reviewer. In most cases you still find that the rater will work for the senior rater and the senior rater will work for the reviewer. If you are a Platoon Sergeant, your Platoon Leader should be your rater, not the First Sergeant. If your First Sergeant rates you, then he should be the senior rater for your Section Chief. Be sure you know who should be your rater. If you are not sure, check with other units like yours. When you sign your name on your NCOER, you are verifying your height, weight and PT test along with part one of the report. If you don't agree with the report, the reviewer will check the block "Nonconcur," attach a comment to the report and you can start your appeal.

Part three of the report is your duty description. This is a very important part of the report because you will be getting promoted in your primary MOS and will be competing against other soldiers in your MOS if you are not working in your MOS. Make sure your primary MOS is in the report or start working in your MOS. If you are working in your secondary MOS, make sure that is also your duty MOS on your NCOER. For daily duties and scope it should show how many soldiers you supervise, what kind of equipment you have, the facilities and the dollar amount. This is very important! Areas of special emphasis is nothing more than your unit goals or the goals the rater has set for you. Also, this should be contained in your counsel statement (which you should get every quarter). So make sure you get them. Your statement should tell you what you should do. When the time comes for you to get your NCOER, comments about the statement should be on the report. Ask for a copy of your statements because they may get lost if you have to appeal your NCOER. Always protect yourself!

Part four is values and NCO responsibilities. If you get all successes and no bullet comments, ask for some. The more you get the better the report. If you get a need improvement, there has to be a comment. You will be rated for:

- Personal commitment, values and Army ethic
- Competencies
- Physical fitness and military bearing
- Leadership
- Training
- Responsibility and accountability

If you get excellent or need improvement in any area, you will need bullet comments; however if you get success ratings, there is no need for bullet comments but you want them anyway. Some raters just won't take the time to write. Tell them you want bullet comments. Success ratings are the same as the old 125 EER. With the new NCOER, it's the bullet comments that give you a good report. So get your bullet comments and make sure they support the rating.

Part five is your overall performance and potential from the senior rater; you will want nothing more than a three. Make sure you get your NCOERs when they are due. Let your rater know a month before time that it will be due. If you need to, write down some bullet comments and give them to him.

MAKE A PHOTO

Each year, send a photo to the Department of the Army. You will be told that you only need one every four or five years, but send one in every year. If you have lost weight, make sure your uniform still fits you (IAW AR 670-1).

SEND A LETTER TO THE BOARD PRESIDENT

A soldier with an announced primary or secondary zone may write the president of the promotion board inviting attention to any matter he feels is important in considering his record. The letter must be received before the convening date.

A letter to the board could put you ahead of your peers because not many soldiers write to the board president. You may ask, "What should I write about?" The best thing to write about is what the board members will be looking for. To find this out you need to go to the promotion section of your Military Personnel Office and ask to see the letter of instructions from the last board. Read it and see if they will give you a copy. If they will not, write down the subjects because they don't change much. If you have the time, try to get the letters from the last two or three boards. That way you can compare them. Some of the subjects may be:

- NCOER, SQT and overall performance
- Level of responsibility
- Trends in efficiency
- Military education
- Professional values
- Range and variety of assignments
- 1SG, Drill Sergeant and recruiting duty
- Derogatory information
- Art 15's and photo
- Physical fitness and overweight
- Medical profiles

There may be more or fewer subjects. Write the letter to the board president. That way they will know you are serious about getting promoted.

Chapter Seven

PROMOTION TO E-7, E-8 AND E-9

ELIGIBILITY FOR PROMOTION

Eligibility for promotion to E-7, E-8 and E-9 is based on date of rank which is announced by Headquarters Department of the Army before the promotion board. General guidance provided by the board states that soldiers competing for promotion to E-7 through E-9 must:

- Meet announced DOR requirements.
- Have the correct TIS when computed for basic pay.
- Be in enlisted status when board convenes.
- Have a high school or GED diploma.
- Not be barred from reenlistment IAW AR 601-280 or QMP.
- Not have an approved retirement before board date.
- Not be eligible for retirement due to a DCSS in effect.
- Must have ANCOC if E-7 with DOR, as of April 1st, 1981 or later.
- Have a minimum of 1 year active duty before board date.

The selection board will be composed of at least five members. The board may be divided into two or more panels. Each panel will have at least three voting members including Officers, Warrant Officers and Senior NCO's. The President of the Board will be a General Officer. Female Officers and enlisted women will be appointed, from time to time, to serve as board members. An officer will be appointed to each board to serve as recorder, but he will not vote. Minority ethnic groups will routinely serve on the board.

Selection will be based on impartial consideration of all soldiers in the announced zone. The best qualified soldiers will be recommended in each MOS for E-7, E-8 and E-9. The board will recommend a specified number of soldiers by MOS from zones of consideration who are the best qualified to meet the needs of the Army. The total number selected by MOS for E-7, E-8 and E-9 is the projected number the Army needs to maintain its authorized-by grade strength at any time.

A separate letter of instruction will prescribe reports to be submitted, largest number to be selected and other administrative details. These documents will be published as enclosures to the letter announcing results of the selection board.

WRITTEN COMMUNICATION

A soldier in the primary or secondary zone for promotion may write the President of the Board inviting attention to any matter he feels is important in considering his record (you will find the address in the document which announced the zone of consideration, which can be found at your personnel section or in the Army Times). These letters may not contain any information on the character, conduct, motives or criticism of any other person. The letter must be received before the board date. If it is not received before the board date, it will not be acknowledged and will not be a basis for promotion consideration.

All letters to the President of the Board are privileged communication to be filed with board proceedings, but will not be included on the soldier's Official Military Personnel File. Letters from third parties, including a soldier's chain of command, are not authorized and cannot be attached as an enclosure.

SELECTION BOARD RESULTS

Headquarters Department of the Army will announce the results of the selection board by a command letter which will include:

- Letter of instruction
- Board members
- Recommended list

Names of soldiers recommended for promotion will be placed in alphabetical order. Sequence numbers for Sergeant Major will be determined by seniority and they will be assigned based on:

- Seniority of date of rank
- Basic active service date, when DOR is the same
- Age (oldest first), when BASD and DOR are the same

Sequence numbers from promotion to grades E-7 and E-8 will be determined by seniority within recommended MOS based on:

- Seniority of date of rank
- Basic active service date, when DOR is the same
- Age (oldest first), when BASD and DOR are the same

Names of soldiers considered for promotion by the board will be listed and an analysis of the board's results by MOS. The analysis provides insight into some of the areas that may influence the board's decision.

EFFECTIVE DATE OF RANK

Promotion to SGM, MSG and SFC will be determined by the Department of the Army on a monthly basis. Effective date of the promotion, for pay purposes, will be the date of the promotion order unless it states otherwise. Soldiers on an MSG or SFC recommendation list who are reclassified prior to promotion, will receive a new sequence number within the new MOS based on their seniority related to order of the soldiers in the new MOS. The new sequence number will be identified by a decimal point. Soldiers who have not been promoted will receive a letter notifying them of their new sequence number. The new number will be based on seniority of all soldiers selected for promotion in a particular MOS (whether promoted or not).

If promotion has already occurred through the new sequence number, the reclassified soldier will be promoted effective the first day of the month following the date of reclassification if the reclassified soldier will be promoted with his contemporaries. If the soldier is reclassified and gets promoted in his old MOS, the orders will remain valid until the soldier's sequence number in the correct MOS is reached and a correct date of rank can be determined. A new promotion order with the correct date of rank, effective date and MOS will be published.

ACCEPTANCE

If a promotable soldier does not decline promotion, it is accepted as of the effective date of the announcing order. Soldiers promoted to grades E-7, E-8 and E-9 will incur a two year service obligation. It will be from the effective date of promotion, before voluntary non-disability retirement, unless they are eligible for retirement by completing 30 years or more of active service or already eligible through prior service or age 55.

The name of the soldier who declined promotion will be removed from the recommended list. The soldier can send a letter of declination through command channels to the MILPO no later than 30 days after the effective date of the promotion given in the order.

APPENDIX

HOW TO VISUALIZE

To imagine is to suppose something. Suppose I did this or I did that. Suppose I was the First Sergeant, suppose I make E-5 next month.

When you imagine something, you create an idea or mental picture in your mind. It is nothing more than a clear image of something you wish to manifest or reveal. Also, we can say thinking is nothing more than imagining something. This is something we do all the time. When we visualize, we are imagining to the point where we actually achieve what we have been thinking about.

Everything we do in life we think about before we do it. Sometimes it happens so fast we find it hard to believe we thought about it first. We can use imagination and visualization to get what we want in life.

Stop and think for a minute about something you wanted very badly and about how you would sit and think about it. Well, what you were doing was visualizing. You were seeing the thing you wanted. You had it on your mind so much that you got it. Now if this is true,

Appendix

why do we have to work? Why can't we just think about what we want and get it? Let me try to explain.

When you visualize, you create a mental picture of what you want. This picture is then transferred to your subconscious mind, which is nothing more than a memory bank. Everything is stored in the subconscious mind. Once it gets information about what you want, it checks through the files to find out what you need to do in order to get it. How is this done? The subconscious mind only works by using pictures; it will flash a picture into your mind and it is nothing more than a flash. Most of the time, this will happen when you are driving your car, in the shower, taking a bath, or feeling happy about something. How many times have you said, or heard someone say, *"Something told me to do that"*? They may say, *"If only I had done what my mind told me to do everything would be all right."* What we do is think about what we want or what we will do and our mind sends it to our subconscious to find out what can be done about it. Our subconscious mind checks through the files for a way to do it and flashes us a picture of what to do. If we pass this up, it will send another and another, until we use what we want to get what we want.

Most of the time we don't get what we want because we come up with something else we want instead; in other words, we give up. The subconscious mind will not check the files unless it is sure that we truly want something. It knows this by how much we think about it. In other words, we have to keep thinking about it. No, it doesn't have to be the only thing we think about, but it has to be what we think about the most. We said before that the subconscious mind will keep flashing pictures to us and the pictures are telling us what to do.

For example, most men, at one time or another, have noticed a woman they wanted to be with. So imagine a man, we'll call him Joe, who sees a woman and is very attracted to her. He wants very much to go out with her. Joe thinks about her so much that his subconscious mind gets hold of it and starts flashing pictures. He may see himself:

- Talking to her on the phone
- Driving her in his car
- Having dinner with her

The pictures are telling him to call her, ask her out for a drive or ask her out for dinner.

When you begin to get flashes about something you want, act on them. Our minds will always flash pictures about what will work, but we have to use them. What happens when the subconscious mind runs out of pictures? It will never run out because it will get more information from you. Do you want to know how? Let's go back to Joe again. He starts getting all these flashes and does nothing about them because he just can't go up and talk to her or call her. His subconscious mind may need to go to another file. So, it flashes a picture of him talking to a friend of hers. He acts on this flash, talks to her friend and finds out that she likes "Rambo" movies. Now, he starts thinking about this and it gets to his subconscious mind. Joe's subconscious mind looks through other files to help find information about talking to her friend. He doesn't think about the movie anymore, just of her. His mind knows that this is something he still wants. A day or two later, while driving home, something tells Joe to stop by the movie. He keeps thinking about this until he stops, finds that a "Rambo" movie is playing and goes inside. Who does he see? The woman he wants to be with, and no one is shy at the movie. If Joe passes up this flash, but keeps thinking about her and his subconscious mind keeps working, time goes by but eventually they get together. They talk and she tells him about the "Rambo" movie she went to see. He says to himself, *"My mind told me to go to that movie and I didn't."*

So you see, when you visualize, you start your subconscious mind working and it flashes you the solution to get what you want. How long it takes you to get what you want depends on how you react to the flashes, and how much information you supply to your subconscious mind so that it can check the files and flash the pictures. The more information you supply, the more flashes you will get, until you pick the ones that get you what you want. Say you want to buy a home, you visualize about it and it is picked up by your subconscious mind. It flashes you a picture of being in a bank getting a loan, but you go to five banks and can't get a loan. Do you give up? No, give more information. To do this, start reading all that you can about buying a home. As this gets into the subconscious mind, it checks the files, starts flashing more pictures and you may get a home where the owner will let you pay him.

Whatever you visualize about, supply all the information you can about what you want. How long will it take to get what you want? I can't say, but it all depends on how much you think about it. Think

Appendix

about it two times a day for about fifteen minutes each time. Give yourself a year or so, if you have been doing this all the time it may not take that long.

To start the visualization process, you will have to think about what you would like to have. Let's say you are an E-4 on the E-5 list, and have been on the list for almost a year. It is now November 1989 and you want to be an E-5 by June of next year. You have to write it down on paper. When you write something down, it's easy for your subconscious mind to pick it up because you are thinking about it while you are writing. When you are writing, there must be a what. The mind will know what you want and where you want it. If you can write it down using colors, do it, because the subconscious mind flashes pictures in color. You may write down something like this:

I, John D. Doe, will be promoted to E-5 in June of 1993, standing on the red carpet, in my Commander's Office.

Always say you **will do** something and not **want to do** something. When you say you will, your subconscious mind believes it more.

When you write this down, keep it with you and read it twice-a-day. Read it like you mean it; that's the only way the mind will work on it for you. Again, you will have to help the mind, so you will need to give it more information about promotion. The way to do this is to read this book, over and over again, your study guide and "How to Sell Yourself" at the end of this appendix. You will start getting flashes. It may be something that you don't think you can do, but act on it. After reading halfway through the book, you may get flashes of:

- You taking a PT test
- You taking an SQT
- You going before the board
- You going to school

And many others.

But whatever it is, act on it. PT test, SQT, board and school are nothing more than points for promotion. How would you feel if you had 760 points, visualized about being promoted, got a flash about taking a PT test, but you didn't do it? Later you learned that the cut-off dropped to 764 points, and you looked on the PT card and saw that all you had to do was run six seconds faster, do two more push-ups, two more sit-ups and you would have gotten promoted?

For visualization to work you have to:

- Write it down
- Supply information
- Act on flashes
- Keep it as your number one thought
- Visualize

Remember earlier when we discussed John getting promoted in June 1993, in his Commander's Office? Well this is what he will visualize happening. He will sit or lie still and just see himself in his Commander's Office, standing on the red carpet, getting promoted. So whatever you write, that is what you want to visualize. It worked for me when I made E-8.

HOW TO SPIT SHINE BOOTS

Why would a soldier want to use his valuable time to spit shine a pair of boots when the regulations say they just have to be evenly shined, an all over "brush shine." Some of the reasons for spit shining boots are:

- Personal pride
- Self-discipline
- Recognition
- Brownie points

Personal pride is the attitude you have about yourself. Do you want to be neat and clean and keep your hair cut and groomed? Are you happy with yourself as you are or would you like to be a better person? Having pride in yourself is nothing more than loving yourself. It's trying to be the you that you want to be. If you are happy being a slob and love yourself that way, your personal pride is being a slob. If that were true, you would not be reading this book because slobs do not get promoted in the Army.

You will also find that the soldiers who have personal pride also have:

- Many friends
- Self-esteem
- Feelings for other soldiers
- Positive images

Self-discipline is continuing to do the things that you know should be done. Initiative is what gets you started in doing what needs to be done. We can put motivation in here also because we said that motivation is the reason for action. Now, let's look at the three together.

Say that you have been trying to make Top Soldier on Guard Duty, but you just can't seem to do it. You know that next week when you have guard duty, the inspector for that day always picks a soldier who has spit shined boots, and you want to be top soldier when you have duty next week. Now you are motivated because you have a reason for action, and the action is to spit shine the boots. The reason is that you want to be the top soldier when you have guard duty next week. Now you go and get your best pair of boots, stop by the Post Exchange and get all the supplies you need to spit shine them. You have just taken the initiative to get it done because you have gotten started. You see, first you got motivated, then you took the initiative to get the job done. You still have a week before you have guard duty, and each day when you get off work you get the boots and start spit shining them, even sometimes during lunch. This is self-discipline, doing what you know has to be done. You keep doing it over and over again, until the time comes for you to have guard duty. Guess what? You make Top Soldier.

That's the way it is when you spit shine your boots. You know they look good and you just keep doing it day after day, without fail. Now comes the recognition because other soldiers will start telling you how good your boots look. The more they tell you this, the more you want to do them. You find yourself doing more for yourself; you may go out and buy a new uniform to go with the boots. You will find that the more you do, the more recognition you get and the more rank you will get. Everyone gets a good feeling about a soldier who has self-discipline. Brownie points are nothing more than recognition. I will tell you now, other soldiers will talk about you, they will call you brown nose or say you are trying to get brownie points, but they are:

- The common, every-day soldier
- Jealous of you
- Soldiers with no pride in themselves
- Soldiers who want to hold you back
- Soldiers who don't know how to do it

So, just look at them and keep going; they may work for you one day!

Now that we know why a few soldiers spit shine their boots, let's see how it is done. First, we will need some supplies. We will need:

- A large can of black Kiwi Shoe Polish
- A small can of neutral Kiwi Shoe Polish
- A large horse hair shoe brush
- A small brush to put polish on the boots
- An old toothbrush
- Some large cosmetic cotton puffs (balls)

Appendix

- Black Kiwi Leather Dye
- Black Kiwi Heel and Sole Dressing
- A cigarette lighter and lighter fuel
- An old towel or rag and some water
- A pair of boot laces

The reason I use Kiwi is because it's good and you can find it at the Post Exchange. Check the Commissary first because it will cost less there. At the Clothing Sales Store you may be able to find a large horse hair shoe brush; get a large one, they are much better. The small brush should also be easy to find, even if you have to buy the small Kiwi Shoe Shining Kit to get one.

Some soldiers use a rag to put the polish on but get a brush. You may have to get the cotton puffs off post if you can't find them in the PX. You need the large ones and they must be 100 percent cotton. You may also use the cotton roll in the blue box, but it's packed too tight and for that reason you will use too much of it. If you feel better wrapping part of a T-shirt around two or three fingers, do so, just be sure the T-shirt is 100 percent cotton. Cotton holds the water better and it will not scratch the boots as you do them, so use 100 percent cotton. You may find the cosmetic cotton balls in different colors but that won't hurt anything.

So now you have all the supplies that you need. All you have to do now is prepare the boots for shining or you may have to take other action. You may have to:

- Get the boots repaired
- Clean off the dirt and mud
- Remove the old shoe polish
- Dye the boots
- Replace the boot laces
- Brush off the boots

Take a look at all of your combat boots, and if you don't have three pairs, try and get three pairs. If you have the boots that are not supposed to be shined, don't try to shine them unless you just want to show others that they can be shined. So, get all of them out and see if you need to put them in the shop to get a heel or sole. Be sure you have all of your boots out because they are all the same.

I have never understood why some soldiers feel that there are:

- Work boots
- Field boots
- Guard boots
- Inspection boots

They are all combat boots. Did you know that some soldiers wear the jungle boots just so they won't have that much to shine? How many Senior NCO's have you seen walking around with leather lust on their boots. Next time you see one, look at his uniform and, most of the time, you will find the boots look better.

Don't be lazy or try to take the easy way out. Leather lust is not self-discipline, it's someone else's discipline. **No short cuts!** If you have dirt or mud on the boots, clean it off and let the boots dry. If the color is fading you will need to dye them. Let them dry very well after you dye them.

I've found that the best way to remove old shoe polish from a pair of boots is to burn it off. You will need:

- An old towel or rag
- The large and small shoe brushes
- Some black shoe polish
- A cigarette lighter and fuel

You should have an old towel to use each time you do your boots. You will need an old rag also, one that you can throw away after you are done. Place the towel across your lap and the boot on your lap and remove the laces. Brush the boots off well with the big brush. Have your rag and lighter close by. Take your small brush and put the black shoe polish all over the boots, but do not brush it off. Once you get this done, take the lighter in one hand and put your other hand all the way down into the boot. Hold the boot up about two inches from the flame of the lighter, moving the lighter back and forth, burning the shoe polish until it turns into a glossy black color. Do only a 2 by 2 inch area, or smaller, at a time. Once your area is glossy black, put the flame out on the lighter and set it down, pick up the rag and wipe off the area you have just burned. The polish will come off onto the rag. Continue doing this until you have gotten it off each boot. You will do the same thing if you have a pair of boots that are hard to shine, or you are spit shining them for the first time and you want to get your base coat started. Do the same thing but don't wipe off the polish.

Always clean the boot laces when you do your boots. This is why you want to buy another pair, so you can rotate them.

Now we will talk about spit shining the boots since we have gotten them prepared. If you did a good job, you will only need:

- The black and neutral shoe polish
- The large and small brushes and the toothbrush
- The cotton balls and some water
- The heel and sole dressing
- The towel for your lap

Lay the towel across your lap and prepare the boot for shining. This time, take your old toothbrush and put some black polish around the top of the sole (this is where the stitches hold the boot onto the sole). Next you will want to put polish all over the boot, then brush it off until they shine.

This is all that the regulations call for, but you want to be better. So put some water in the top of the can of neutral polish. Get a cotton ball and hold it with two or three fingers, whichever is best for you. Make sure you have a firm hold on the cotton ball because this is what you will use to shine the boots. You will only need about one-fourth of the top filled with water. Holding the cotton ball, put it into the water for about a second and take it out. With your other hand over the top as if you were going to chop it, press the cotton ball into the palm of the hand, over the top, pressing it into the palm so that the water will run back into the top. At the same time you can mold the cotton ball around your fingertips like you want it.

Now that you have most of the water out of the cotton ball, rub it around in your black shoe polish until the bottom is covered and start rubbing it onto the boot in a circular motion. From time-to-time, take one of the fingers that is not holding the cotton ball and dip it into the water; pat it on the part of the boot that you are shining. Keep rubbing it onto the boot until you see it shining. Put more polish on the cotton ball, and more water on the boot. You can also put the water on the part that you are going to shine before you put the polish on and start rubbing it.

If you rub too hard, or if the shoe polish starts coming off, use less pressure and more polish and water. Also, if the polish is too old it will come off. If you find that it is hard to get the boot to shine, you can burn some polish onto it as if you were going to take the old polish off, but don't wipe it off. When it cools off, start with the cotton balls again until you get it to shine. You have to keep the shoe polish moist, but don't use too much water. Keep doing it until you get it to shine.

Once you get the boots to shine the way you want them, go over them again the same way with the neutral shoe polish. This is what gives them the glassy shine. You know you have it right when other soldiers start asking you what you have on your boots. Don't forget to do the tongue of the boot.

This is how I used to make Top Soldier on Guard Duty, because most of the soldiers who were trying to make top soldier just shined the heels and toes of their boots.

If the boots seem like they will not shine, just keep trying and they will. The shoe polish has to build up on them first. Use the lighter and burn the polish until you can get them to start shining. Once you get them to shine, then go over the heel and sole with the heel and sole dressing.

If you spit shine your boots, do it all the time because the other soldiers will be waiting for you not to do it. Who in your unit spit shines their boots all the time? Do you check their boots when you see them?

Some people like to see other people kept down; they don't want to see you move up. Most of the time it's because they don't know what to do for themselves to improve, or they just don't have the willpower that you have. Let them talk, but when you see the information, look at your boots and look at theirs. Be proud, that's what it's all about! Don't give up.

Appendix

HOW TO SET UP YOUR WALL LOCKER

Your wall locker is you. Did you ever notice how an inspector looks at a displayed locker that is messy, turns around and looks at the soldiers? Why is this? Could it be:

- Organized soldiers seek responsibility?
- Are they paying attention to detail?
- Neatness and cleanliness?
- Personality trademarks?

An organized soldier seeks responsibility and is one that keeps things in their rightful place. He wears his uniform with pride and according to the regulation. His equipment and personal gear are clean and serviceable at all times. His room and wall locker are well organized. Everything in his wall locker is according to the wall locker display, if he has one; if not, his locker can be used as a role model to draw one up.

Nine times out of ten, an organized soldier will have an organized section, company or whatever group of people he is in charge of. High ranking NCO's in higher headquarters don't get to see the soldier as much as they would like to because of other commitments, so they judge the soldier by:

- His military bearing
- Appearance of his room, wall locker and equipment
- His military knowledge
- How he feels about himself, his unit and the Army

All of these things are checked when you have a command inspection or higher; your Commander and First Sergeant will check the same things to prepare you for those inspections. Do you know how you can always be prepared for an inspection seven days a week, twenty-four hours a day? The only way is to keep it prepared at all times.

Are they paying attention to detail is what the inspector wants to know. This is why you will see them with the wall locker display, or looking at the first locker very closely to see how it is set up. Once they have a display or see how the locker is set up, they will check and see if the others are the same. Some inspectors feel that if you can't set up your locker the way it is on display, you may not be responsible enough to have your soldiers where they should be if you were in a combat zone. How would you feel about the soldier if you were the inspector? If you can do it today, you may inspect it later. Neatness and cleanliness are something you should have brought into the Army with you.

You should be neat and clean at all times and you should make sure that your soldiers are the same way. There may be a time when you find a soldier who doesn't keep himself clean. You should:

- Tell him about it
- Tell your Chain of Command
- Tell the Medic
- Write him a letter
- Give him cleaning supplies
- Have one of his friends tell him
- Help him do it

Your uniform should be clean and pressed when you put it on in the morning, along with your footwear and headgear. Most likely, if you are neat and clean, your work area, car, home and room will be clean. Have pride in yourself and be neat and clean; it's easy!

Personality trademarks are the things that you have that reflect on your personality. They could be:

- Pictures you have displayed
- Dirty clothes
- Unauthorized weapons or drugs
- Unserviceable military or civilian clothes
- Boots and shoes that are not cleaned and shined

Look around at your things. What do they tell about your personality? Could they give off the wrong signals? You be the inspector; what do you think? If you are not sure, have one of your friends check you out. Again I will say, your wall locker is you. We can say your wall locker is a home that you are buying; it is about five years old and just waiting for someone to move in. Some of the things you may do before and after buying the home are:

Appendix 93

- Inspect the home inside and outside
- Make plans for improvements
- Get the materials for repair
- Clean it up
- Get it organized

You can do the same things with your wall locker. First, you will need to take everything out of the locker and put it as far away as you can. If the locker is movable, move it away from the wall so you can see all around it and get behind it. Now get a pen and a note pad and do an inspection on the locker; you will do it the same way as we do the home.

Inspect the home (locker) on the outside by standing back about two or three feet and getting an overall view of it. Look at it all the way around, go back to the front, stand about a foot from it and start the inspection. Check it from top-to-bottom, from side-to-side, move around to the back and do the same thing. Write down any and everything that you see wrong with it. This could be:

- Screws, nuts and bolts missing
- Lock is unserviceable
- Doors are not flush
- Top, bottom or sides are not secure
- Locker needs painting

Check your locker on the inside. Now that the outside is done, you may find:

- Mirror, clothes rack or towel rack missing
- The inside needs painting
- Drawers missing hardware
- Top or bottom shelf not secured

Since you have inspected the locker on the outside and inside, and you have written what's wrong or missing, you can make a list of the things you will need to make improvements. You may need:

- A wall locker display
- A room SOP
- Cleaning supplies

- Parts for repairs
- Paint or wood putty

Make plans for improvements according to how much work needs to be done on the locker. If you have to paint it, you may want to do it just before you go:

- Out to field training
- On a long weekend

Most of the time you will find that you will be able to make your improvements during the weekend. You will want to plan your inspection a-day-to-a-week, depending on how long it will take you to get the things you will need for the improvements.

Once you have made your plans you need to get your supplies. You can check:

- With your supervisor
- In the supply room
- At the PDO
- With the Platoon Sergeant
- With the First Sergeant
- Soldiers who are using lockers for equipment storage

Once you get the parts, repair the wall locker. If you get a replacement, inspect it the same way you did your old locker.

Cleaning your locker is the next thing that has to be done. You may need:

- Hot soapy water
- Brasso
- Liquid Gold (wood locker)
- Car wax (metal locker)
- Contact paper (wood locker)

After you clean the locker, go over it again with clean, hot water to remove the soap film. When it is dry, do the same thing with Liquid Gold; this will give new life to the wood finish. If you have a removable wood clothes rack, remove it, sand it down with sandpaper and put two or three coats of clear varnish on it; let it dry and put it

Appendix

back into the locker. Or you can paint the wood clothes rack a glossy white color. If there is a metal clothes rack, remove it, Brasso it and put it back. Try to get a chrome rack for the wood or metal wall locker.

Now for the metal locker you will clean it the same way with hot, soapy water, letting it dry and going over it again with hot, clean water. If you have to paint the locker, paint it with the same color paint, or one that is approved by the First Sergeant or Commander. If you have to paint it, wait a month or more before you use the car wax on it. If you do not have to paint the locker, after you clean it wax it down with the car wax. Do this until you can see yourself on the outside and inside. Afterwards, do this about once every month. Next, take Brasso and put a good shine on the clothes and towel racks if there are any. Now stand back and see how good your locker looks. It should look brand new! Good job!

Getting it organized is all you have to do now and you will be ready for the President, if he comes by. Get your wall locker display and get started. Also, it will help to have:

- The same kind of hangers
- All cleaned and pressed clothes
- Roll-on or solid deodorant

Look at your clothes hangers. Are they all the same kind and from the same place? If not, you will need the same kind. You don't have to go out and buy new ones, but if you do get the same kind. If you get wood, they should all be wood; if you get red, they should all be red. You want the same kind because your clothes will all hang the same way and will be even across the top and bottom.

Before you start hanging them, be sure they are clean and pressed. Look at your display and see how they are to be placed in the locker. Check out your room SOP also; it may have something pertaining to your wall locker. Now you want to get all of your things hung on the hangers, but not in the locker as yet. If your display says that all of the buttons must be buttoned, make sure they are buttoned.

Some of the places that inspectors check are:

- On the fly of the BDU's
- The back pockets of the trousers
- The buttons on the front, inside and bottom of the coat

- Top buttons on shirts and jackets
- Buttons on your civilian clothes

When you put your trousers on the hanger, make sure all of your buttons are buttoned and the front end (fly of the trousers) is facing you. If you do it this way, when the inspector checks the buttons on the trousers they will be easy to check and will all be the same.

Be sure to do your civilian trousers the same way. All trousers should have a shirt or jacket over them. Not only does this save room in your locker, but it makes it look neater. You want to button all of the buttons on your jackets (BDU shirts), place them over the trousers and begin hanging them up. Don't forget that the front end of the trousers should be facing you and the top, round, open end of the hanger should be facing the back of the locker when hung.

Be very careful when you hang your clothes because your clothes rack has a very high shine or paint job. After you get them in the locker according to the display, you can work on the eye appearance. Once all your things are hung in the locker you need to get the hangers spaced evenly. To do this, get a ruler and space the hangers two or three inches apart, depending on how much room you have available. After this is done, you can line up your unit patches so that they are in a straight line. Move the sleeve up or down to align it with the others. The collars should be about even, the patches should be on line and the bottoms about the same. Now you see why you need the same kind of hangers.

Looking over your display, place the other things in the locker as they are on the display. Don't cover your shelves with anything unless it says so on the display. It would look good, but don't forget that the inspector wants to see if it is like the display. If you can cover the shelves, use the brown towels; they look more military.

You do not want to display anything new in your locker. If you have new underwear, wash them before you display them. Use your soap once before you put it on display. Do the same thing with your shoe polish. Remember, nothing new!

Don't use gold plated brass; inspectors may think you like to take the easy way out. Don't even have it in your locker.

Make sure you have all your rolled and folded up things the same size.

Make sure you have a good serviceable toothbrush, but not a new one; not even a new tube of toothpaste.

Once you are done, it's time to check it out. You can:

- Check it out by the display
- Have your boss inspect it
- Have the Platoon Sergeant check it

After you get your locker set up, have it inspected. That's the way you double check yourself. You know it's good but others will:

- Tell you you did a good job
- Give you more suggestions
- Tell others how good it looks
- Have the other soldiers do the same
- Have you inspect other lockers
- Show your locker to higher ranking soldiers
- Use your locker to draw up the display
- Remember it when it's time for your promotion

A soldier that lives in the barracks has an advantage over the soldier who lives off post for promotion because:

- He is seen more by High Ranking Officers and NCO's
- He has a room and wall locker to be inspected
- His Chain of Command knows how he lives

Try to keep the things on your display in your wall locker. The less you have in your locker, the less you will have to be inspected. The best way to do this is to keep only the things you would take with you if you were getting out of the Army tomorrow. Well, all you have to do now is to keep it looking good. You did a good job. Good luck!

HOW TO SELL YOURSELF

Before you go to the promotion board, be prepared to sell yourself. What you say about yourself when asked is very important because it:

- Can get you points later
- Can pull voting members to your side
- Can put you in control
- Will help you relax
- Will give you more confidence

Once you start selling yourself to the board members you can get points faster, because if you do a good job some of the voting members know that they will recommend you for promotion. As you talk, some of the board members will start pulling for you. They will want you to do well. You can take control by using eye-to-eye contact and speaking so that they can hear you. It will help you to relax because you will see the members pulling for you, and that will give you more confidence in yourself and what you are saying. If you can sell yourself well, you can get recommended for promotion and pick up 200 points at the same time.

Before you can sell yourself at the promotion board you will need to know:

- How to talk to all the members
- How to use eye-to-eye contact
- Why you entered the Army
- Things you have done in the Army
- Things you will do when promoted
- How to write your presentation

So far we have been talking about getting promotion points; now we are going to talk about getting a recommendation for promotion along with another 200 promotion points. LET'S START!

How to Talk to All the Members

When the President of the Board asks you to tell something about yourself, talk to all the board members except the recorder because the

Appendix

recorder is not a voting member. To explain how this is done I will invent names for the president and voting members.

- Board President   SGM Tank
- Voting Member   1SG Stinger
- Voting Member   1SG Hawk
- Voting Member   1SG Standards
- Voting Member   1SG Run

In this case, the Sergeant Major will not vote unless there is a tie. The way they are seated is as follows: the Sergeant Major at the center of the long table with 1SG Hawk at his far right and 1SG Stinger on his immediate right, 1SG Run is at his far left with 1SG Standards at his immediate left. They will be sitting behind a long table, which is about five feet in front and center of the chair you are sitting in. You will want to talk loud enough so that 1SG Hawk can hear you on the far right and 1SG Run can hear you on the far left.

Now picture someone standing about two feet behind each voting member. When you are talking to the members, talk as if you are talking to the soldier standing behind the voting members also. When you are talking to only one member, do the same thing. If you do this all the members will be able to hear you. Whatever you do, don't start talking low because you don't know the answer to a question. Try to keep the same voice tone and level. Sometimes, if a soldier is doing well, a member may cut him only one point if he misses one of the questions.

How to Use Eye-to-Eye Contact

Now that you are speaking up and have their attention, you are in control. But do you really have their attention? No! So how do you get it? The Sergeant Major will ask you to tell the members of the board something about yourself. You can say, *"Sergeant Major Tank and members of the board . . ."* If you say it that way, the Sergeant Major knows that you are talking to him, and the members know you are also talking to them. The recorder is a member of the board, but not a voting member. The SGM may not be a voting member unless there is a tie, at which time he will break the tie.

I'm sure you have been walking along and someone spoke to you by using your rank and name. How did it make you feel? If nothing else, it got your attention and that's what you want to do here. Get their attention. As you do this, you need eye-to-eye contact or face-to-face contact. Use the method that is best for you. You may even look at the board members' foreheads.

Everything you do at the board, you want to do in a military manner. As you start talking, you want to look at the SGM first, then 1SG Hawk, 1SG Stinger, SGM Tank, 1SG Standards and 1SG Run. This is what you will say and do. Looking at the SGM, address him, *"Sergeant Major Tank,"* then look to your far left and say, *"1SG Hawk,"* then look at the next 1SG and say *"1SG Stinger,"* look at the SGM again, but say nothing to him, then look at the 1SG on his left and say, *"1SG Standards,"* and finally the 1SG next to him and say, *"1SG Run."* You have gotten the attention of all the voting members. You are letting them know that you are talking to them. You are speaking up and you have their attention. Now what will you say?

Why You Entered the Army

To start your presentation, you want to tell them why you entered the Army. Sure, you can say, *"I joined the Army in May 1985, and had Basic Training at Fort Bliss, Texas."* But will that keep their attention? I don't think so. You see, you have called all the members by their rank and name and you are speaking so that each of them can hear you. The thing you want to do now is get some smiles or maybe even a laugh. This will pull them more to your side; you don't want them to start writing until your presentation is over and the first voting member starts asking you questions.

So, what do you say? That is something you will have to decide. But what if you said something like, *"Before I joined the Army in May 1985, I was working at Bob's Bar-B-Que, where I had been laid off three times within a year and my wife told me that if I got laid off again she would lay me off forever and go home to her mother. So I was glad to be at Fort Bliss, knowing that after training I would be with my wife again and wouldn't have to worry about being laid off again."*

Or you can say something like, *"One day, while I was sitting at home watching TV, an old friend came by to see me. At the time I didn't know he was in the Army, but before he left he talked me into joining also. What he didn't tell me about what getting up at 04:00 in the morning and marching and standing in the rain for hours saluting a tree because I didn't know what a Captain was when the Commander asked me his rank."*

What if a female soldier, said something like, *"All my life I've been a tom-boy and that's one of the reasons I wanted to join the Army. Before Basic Training was over, I wished I had been just a plain girl all of my life. It was hard for me, but fair, because we were all treated like soldiers."*

What will you say? I'm sure you can come up with something to make them smile.

Things You Have Done in the Army

You have just passed the main part of your presentation. You have all the members' attention and they are all looking at you. Now it's time to sell yourself. Tell them what you have been doing since you have been in the Army, beginning with Basic Training. Tell them about everything that you did well, any leadership position you had and any unit you were in, including the one you are in at the time of the board. Check all of the awards in your records because that's what the members will be looking for when they check your records. Talk about what is in your records. There may not be anything about your leadership position in Basic Training or Advance Training (MOS). Some are:

- Squad Leader
- Platoon Guide
- Class Leader

You may or may not have a letter in your records about those positions, if not, be sure to tell about them. Talk about your awards in the order you received them and tell them about the main reason for each award. Also, tell them about any letters or Certificates of Achievement you may have.

If for some reason you had an Article 15, tell about it. If it's not in your records, it may still be recorded on your Form 2A if you were flagged. OK, let's say you had an Article 15 and some awards and you are talking to the board members. You may say something like this, *"During Basic Training at Fort Bliss, Texas, I was a Squad Leader for the Third Squad, First Platoon. When I went to AIT at the same post, I was picked to be Class Leader and graduated number* two in my class. *After a week of leave, I was sent to C Battery, 1st Battalion, 7th ADA. There I received an Army Achievement Medal as an End of Tour Award. Also, I was given an Article 15 because of a misunderstanding in an order I was given. I have been in this Battalion since June of 1989 and I received an Army Commendation because I had the best Arms Room on post during the post IG Inspection. Because I scored 300 on the Physical Training Test, I got a Certificate of Achievement from Lieutenant Colonel Red."*

Things You Will Do When Promoted

So far you are looking good. Now you want to tell them what you will do **when** you get promoted. Not **if** you get promoted. This will let them know that you are sure of yourself. You need to know:

- What is hot at the time
- What are the Battalion goals.
- What is needed Army wide

To find out what is hot at the time, there are things you can do:

- You can look at the news on TV
- You can read the *Army Times*
- You can check with your PAC, NCO and 1SG
- You can talk to other soldiers about what's going on

If you don't know what the Battalion Goals are, you can ask your First Sergeant, Platoon Sergeant or your Sergeant Major. Some units have a monthly newsletter, which could contain the goals of the unit as well as those of the Battalion. Some goals may be:

- To get more licensed drivers
- To give better soldier care

Once you get all of this information, all you need to do is say something about it. Then the members can relate to everything you talk about. You may say something like this, *"I understand our field training will be cut in half because of the budget cuts throughout the Army, but this will give me a chance to train my soldiers for tomorrow's wars which will be needed Army wide. Also, I will have more*

Appendix 103

time to train my soldiers and to ensure that they all have their driver's licenses and licenses to operate the generators. I want to show my soldiers that, not only do I care about how hard they work, but also about them. Each month I will visit one or more of my soldiers at their homes, just to talk to them and their families about anything they want to talk about. If the soldier lives on post, then I will take him to my home or some other place where we can talk. My wife supports me 100 percent and she understands my mission. I can give you one guarantee, that is you will never regret recommending me for promotion."

How to Write Your Presentation

After you get all your information, you want to write it down and read it until you know it by heart. Also, you can put it on tape and play it when you are in your room, home or car. So you have all the information. It may look like this. *"Before I joined the Army, I was working at Bob's Bar-B-Que, where I was laid off three times within a year. My wife told me that if I got laid off once more she would lay me off for good and go home to her mother. So I joined the Army in May 1985. I had Basic Training at Fort Bliss, Texas, where I was the Squad Leader for the Third Squad, First Platoon. When I went to AIT at the same post, I was picked to be Class Leader and graduated number two in my class. After one week of leave, I was sent to C Battery, 1st Battalion, 7th ADA where I received an Army Achievement Medal as an End of Tour Award. Also, I was given an Article 15 because of a misunderstanding in an order I was given. I have been in this Battalion since June of 1989 and I received an Army Commendation Medal because I had the best Arms Room on post, 'in writing,' by the post IG Team. When I scored 300 on the PT Test, I got a Certificate of Achievement from Lieutenant Colonel Red. I understand our field training will be cut in half because of the budget cuts throughout the Army, but this will give me a chance to train my soldiers for tomorrow's wars which may become a big issue later. I will ensure that my soldiers have their driver's licenses for the trucks and generators. I will show all of my soldiers that I care about more than just the work they do. I will visit with them each month to show my support and to show them that I care. My wife supports me all the way and understands my mission. I can guarantee you all one thing and that is you will never regret recommending me for promotion to the next highest grade."*

There you have it. Once you write it down, read it and make some changes if you feel you must. Don't forget you can put it on tape also. When you are giving your presentation, you want to look at all of the members. You don't have to say their names anymore because you did that to get their attention, and you have it now. As you talk, look at one member at a time, each for two or three seconds. When you are done with your presentation, look at the President of the Board; he may ask you a question or two, just to relax you more. Then one of the First Sergeants will start asking you questions. When a voting member asks you a question, you should:

- Call him by his rank and name
- Repeat the question
- Answer the question

It may go like this. *"First Sergeant Run, the Battalion Commander is Lieutenant Colonel Red."* After all the voting members have asked their questions, the President may ask them if they have any more questions for you. If they all say no, he may ask you if you would like to ask the members anything. At this time you can thank them. You will do this the same way that you started your presentation, by calling each by their rank and name. Again, start with the President of the Board.

When the president says you are done, jump to attention as if Colonel Red had walked into the room. March to the table, salute and hold it until it is returned, make your facing movement and march out the door.

Don't talk to anyone about what you did or said during the board.

The best times to appear before the board are first, last or the first person before or after lunch.

What do you get when you do a good job selling yourself? You will get a recommendation and 200 points from each voting member. That should be your goal. Do it!

Now that you have the knowledge, it's no good unless you use it.

Appendix

CHAIN OF COMMAND

Commander in Chief   _____

Secretary of Defense   _____

Secretary of the Army   _____

Chairman, JCS   _____

Chief of Staff, Army   _____

MACOM Commander   _____

Post Commander   _____

Brigade Commander   _____

Battalion Commander   _____

Company Commander   _____

Platoon Leader   _____

SUPPORT CHAIN OF COMMAND

Sergeant Major of the Army   _____

MACOM Command Sergeant Major   _____

Post Command Sergeant Major   _____

Brigade Command Sergeant Major   _____

BN Command Sergeant Major   _____

First Sergeant   _____

Platoon Sergeant   _____

Section Chief   _____

Squad Leader   _____

OTHER VIP'S IN CHAIN OF COMMAND

INDEX

A
Acceptance, 79
Accessories, 24
Accomplishments, 1, 101-102
Alcohol and drug prevention
  and control program, 45
Appearance, 53, 85-90, 92
Appointments, 7
Attitude, 3-5, 25, 56
Authority, 12
Awards, 52

B
Bearing, 21-25, 53
Board members, 58-61, 68
Boots, 85-90

C
Chain of command, 105-106
Clothes, 18-19
Commander, 46
Communication, 98-99
Conversational skills, 54

D
Decorations, 52
Departure, 69-79
Dress, 21-25, 66-68
Duty performance, 49

E
E-7, 71-79
E-8, 71-79
E-9, 71-79
Education:
  civilian, 44-45, 51
  military, 49-50
Eligibility, 35-45, 76-77
Equipment, 16-21
Expression, 68-69
Eye contact, 10, 99-100

F
Faith, 3
Fitness text, 27

G
Goals, 5
Guard duty, 9-15

H
Head gear, 23
Height, 41

I
Inspections, 11
Instructor, 30

K
Knowledge, 54-56

L
Leadership, 44
Letter of recommendation, 56
Learning, 31-33
Letter, 75
Locker, 21

M
Military knowledge, 12-15
Motivation, 2-5, 100-101

N
NCOER, 73-74

O
Officers, 5-6
Oral expression, 54
Orders, 5-7
Organization, 91

P
Peers, 1
Performance statements, 8-9
Photo, 74
Physical fitness test, 39-40
Pre-boards, 64-65
Private E-2, 36
Private E-3, 36
Problem-solving, 32-33
Promotion
 board, 11, 52-56, 57-65
 eligibility for, 76-77
 guide, 63
 points, 42, 46-56
 recommendations, 42-43
Promotions, 1

Q
Quarters, 20
Questions, 70

R
Rank, 78-79
Records, 62-63, 71-72
Respect, 69
Results, 102-103

S
Schooling, 44
Security clearances, 42
Selection board, 77-78
Self-confidence, 3, 53
Self-knowledge, 2
Self-promotion, 98-106
Selling, 54, 98
Sergeant E-5, 37-38
Sergeant E-6, 37-38
Sergeant E-7, 38-39, 71-75
Sergeant E-8, 38-39, 71-75
Sergeant Major, 38-39
Skill Qualification Test, 26-30, 48-49
Soldier care, 33-33
Specialist E-4, 36-37
Spit shining, 85-90
Storage, 19-20
Study guide, 12-15, 61-62
Superiors, 10-11

T
Time, 7-8
Time in grade, 36-39
Time in service, 36-39
Training, 51-52

U
Uniform, 21-24, 63-64, 92

V
Visualizing, 5, 80-84

Index

W
Waivers, 42
Wall locker, 91-97
Weapons, 16-17, 30-31
Weapons qualifications, 41
Weight, 28, 41
Writing, 77, 103-104

ABOUT THE AUTHOR

How long Sergeant Walker stayed in the Army and what war he was in are not important. It's not important how many awards he has or how many times he maxed the physical fitness test.

What is important is that Sergeant Walker wrote this book with knowledge gained from over 12 years experience as a voting promotion board member. He also served as the President of many pre-boards. He wrote a study guide for his battalion in 1982 when he was stationed in Germany. Sergeant Walker was Soldier of the Quarter when he was an E-5 and made Soldier of the Year as an E-7. He has served as a Section Chief, Platoon Sergeant and First Sergeant. Sergeant Walker has helped many soldiers prepare for promotion and most of them max the board.

Sergeant Walker is a Soldier's Soldier. He was always hard on NCO's because he believed a sorry NCO will create sorry soldiers.

Master Sergeant Walker retired from the Army but not the soldiers. This book is his way of staying in touch.

CAREER RESOURCES

Contact Impact Publications to receive a free copy of their latest comprehensive and annotated catalog of over 1,400 career resources (books, subscriptions, training programs, videos, audiocassettes, computer software).

The following career resources are available directly from Impact Publications. Complete the following form or list the titles, include postage (see formula at the end), enclose payment, and send your order to:

IMPACT PUBLICATIONS
9104-N Manassas Drive
Manassas Park, VA 22111
Tel. 703/361-7300
FAX 703/335-9486

Orders from individuals must be prepaid by check, moneyorder, Visa or MasterCard number. We accept telephone and FAX orders with a Visa or MasterCard number.

Qty.	TITLES	Price	TOTAL
MILITARY			
___	Army Officer's Guide	$17.95	_____
___	Beyond the Uniform	$12.95	_____
___	Civilian Career Guide	$12.95	_____
___	Combat Leader's Field Guide	$10.95	_____
___	Combat Service Support Guide	$14.95	_____
___	Does Your Resume Wear Combat Boots?	$9.95	_____
___	Enlisted Soldier's Guide	$12.95	_____
___	Guide To Effective Military Writing	$14.95	_____
___	Guide To Military Installations	$17.95	_____
___	Job Search: Marketing Your Military Experience	$14.95	_____

___ NCO Guide $17.95 _____
___ Re-Entry $13.95 _____
___ Retiring From the Military $22.95 _____
___ Service Etiquette $29.95 _____
___ Soldier's Study Guide $12.95 _____
___ Today's Military Wife $14.95 _____
___ Up or Out? How To Get Promoted as the
    Army Draws Down $13.95 _____
___ Woman's Guide To Military Service $9.95   _____
___ Young Person's Guide To the Military $9.95   _____

BEST JOBS AND EMPLOYERS FOR THE 90s

___ 100 Best Careers For the Year 2000 $14.95 _____
___ 100 Best Jobs For the 1990s and Beyond $19.95 _____
___ 101 Careers $12.95 _____
___ American Almanac of Jobs and Salaries $15.95 _____
___ America's 50 Fastest Growing Jobs $9.95   _____
___ America's Fastest Growing Employers $14.95 _____
___ Best Jobs For the 1990s and Into the 21st Century $12.95 _____
___ Hoover's Handbook of American Business (annual) $24.95 _____
___ Hoover's Handbook of World Business (annual) $21.95 _____
___ Job Seeker's Guide To 1000 Top Employers $22.95 _____
___ Jobs! What They Are, Where They Are, What They Pay $13.95 _____
___ Jobs 1993 $15.95 _____
___ Jobs Rated Almanac $15.95 _____
___ New Emerging Careers $14.95 _____
___ Top Professions $10.95 _____
___ Where the Jobs Are $9.95   _____

KEY DIRECTORIES

___ Career Training Sourcebook $24.95 _____
___ Careers Encyclopedia $39.95 _____
___ Dictionary of Occupational Titles $39.95 _____
___ Directory of Executive Recruiters (annual) $39.95 _____
___ Directory of Outplacement Firms $74.95 _____
___ Directory of Special Programs For Minority
    Group Members $31.95 _____
___ Encyclopedia of Careers and Vocational Guidance $129.95 _____
___ Enhanced Guide For Occupational Exploration $29.95 _____
___ Government Directory of Addresses and
    Telephone Numbers $95.00 _____
___ Internships (annual) $28.95 _____
___ Job Bank Guide To Employment Services (annual) $149.95 _____
___ Job Hunter's Sourcebook $59.95 _____
___ Moving and Relocation Directory $125.00 _____
___ National Directory of Addresses & Telephone Numbers $89.95 _____
___ National Job Bank (annual) $239.95 _____
___ National Trade and Professional Associations $79.95 _____
___ Minority Organizations $49.95 _____
___ Occupational Outlook Handbook $21.95 _____
___ Professional Careers Sourcebook $79.95 _____

JOB SEARCH STRATEGIES AND TACTICS

___ But What If I Don't Want To Go To College	$10.95	_____
___ Career Planning and Development For College Students and Recent Graduates	$17.95	_____
___ Careering and Re-Careering For the 1990s	$13.95	_____
___ Complete Job Search Handbook	$12.95	_____
___ Dynamite Tele-Search	$10.95	_____
___ Get the Right Job in 60 Days or Less	$12.95	_____
___ Go Hire Yourself an Employer	$9.95	_____
___ Guerrilla Tactics in the New Job Market	$5.99	_____
___ How To Get Interviews From Classified Job Ads	$14.95	_____
___ Joyce Lain Kennedy's Career Book	$29.95	_____
___ Knock 'Em Dead	$19.95	_____
___ Perfect Job Search	$12.95	_____
___ Professional's Job Finder	$15.95	_____
___ Right Place At the Right Time	$11.95	_____
___ Rites of Passage At $100,000+	$29.95	_____
___ Super Job Search	$22.95	_____
___ Take Charge of Your Career	$10.95	_____
___ Who's Hiring Who	$9.95	_____
___ Work in the New Economy	$14.95	_____

CITY AND STATE JOB FINDERS

___ California: Where To Work, Where To Live	$9.95	_____
___ Jobs in Washington, DC	$11.95	_____
___ L.A. Job Market Handbook	$15.95	_____

How To Get a Job In . . .

___ Atlanta	$15.95	_____
___ Boston	$15.95	_____
___ Chicago	$15.95	_____
___ Dallas/Fort Worth	$15.95	_____
___ Houston	$15.95	_____
___ New York	$15.95	_____
___ San Francisco	$15.95	_____
___ Seattle/Portland	$15.95	_____
___ Southern California	$15.95	_____
___ Washington, DC	$15.95	_____

Bob Adams' Job Banks to:

___ Atlanta	$15.95	_____
___ Boston	$15.95	_____
___ Chicago	$15.95	_____
___ Dallas/Fort Worth	$15.95	_____
___ Denver	$15.95	_____
___ Detroit	$15.95	_____
___ Florida	$15.95	_____
___ Houston	$15.95	_____

___ Los Angeles	$15.95	_____
___ Minneapolis	$15.95	_____
___ New York	$15.95	_____
___ Ohio	$15.95	_____
___ Philadelphia	$15.95	_____
___ Phoenix	$15.95	_____
___ San Francisco	$15.95	_____
___ Seattle	$15.95	_____
___ Washington, DC	$15.95	_____

ALTERNATIVE JOBS AND CAREERS

___ Adventure Careers	$9.95	_____
___ Advertising Career Directory	$17.95	_____
___ Book Publishing Career Directory	$17.95	_____
___ Business and Finance Career Directory	$17.95	_____
___ But What If I Don't Want To Go To College?	$10.95	_____
___ Career Opportunities in Advertising and Public Relations	$27.95	_____
___ Career Opportunities in Art	$27.95	_____
___ Career Opportunities in the Music Industry	$27.95	_____
___ Career Opportunities in the Sports Industry	$27.95	_____
___ Career Opportunities in TV, Cable, and Video	$27.95	_____
___ Career Opportunities in Theater and Performing Arts	$27.95	_____
___ Career Opportunities in Writing	$27.95	_____
___ Careers For Animal Lovers	$12.95	_____
___ Careers For Bookworms	$12.95	_____
___ Careers For Foreign Language Speakers	$12.95	_____
___ Careers For Good Samaritans	$12.95	_____
___ Careers For Gourmets	$12.95	_____
___ Careers For Nature Lovers	$12.95	_____
___ Careers For Numbers Crunchers	$12.95	_____
___ Careers For Sports Nuts	$12.95	_____
___ Careers For Travel Buffs	$12.95	_____
___ Careers in Computers	$16.95	_____
___ Careers in Education	$16.95	_____
___ Careers in Health Care	$16.95	_____
___ Careers in High Tech	$16.95	_____
___ Careers in Law	$16.95	_____
___ Careers in Medicine	$16.95	_____
___ Careers in Mental Health	$10.95	_____
___ Careers in the Outdoors	$12.95	_____
___ Encyclopedia of Career Choices For the 1990s	$19.95	_____
___ Environmental Career Guide	$14.95	_____
___ Environmental Jobs For Scientists and Engineers	$14.95	_____
___ Health Care Job Explosion	$14.95	_____
___ Healthcare Career Directory	$17.95	_____
___ Magazine Publishing Career Directory	$17.95	_____
___ Marketing and Sales Career Directory	$17.95	_____
___ Newspaper Publishing Career Directory	$17.95	_____
___ Opportunities in Accounting	$13.95	_____
___ Opportunities in Advertising	$13.95	_____
___ Opportunities in Beauty Culture	$13.95	_____
___ Opportunities in Biological Sciences	$13.95	_____

Career Resources 115

___ Opportunities in Chemistry	$13.95	_____
___ Opportunities in Civil Engineering	$13.95	_____
___ Opportunities in Computer Science	$13.95	_____
___ Opportunities in Counseling & Development	$13.95	_____
___ Opportunities in Dental Care	$13.95	_____
___ Opportunities in Electronic & Electrical Engineering	$13.95	_____
___ Opportunities in Environmental Careers	$13.95	_____
___ Opportunities in Financial Career	$13.95	_____
___ Opportunities in Fitness	$13.95	_____
___ Opportunities in Gerontology	$13.95	_____
___ Opportunities in Health & Medical Careers	$13.95	_____
___ Opportunities in Journalism	$13.95	_____
___ Opportunities in Laser Technology	$13.95	_____
___ Opportunities in Law	$13.95	_____
___ Opportunities in Marketing	$13.95	_____
___ Opportunities in Medical Technology	$13.95	_____
___ Opportunities in Microelectronics	$13.95	_____
___ Opportunities in Nursing	$13.95	_____
___ Opportunities in Paralegal Careers	$13.95	_____
___ Opportunities in Pharmacy	$13.95	_____
___ Opportunities in Psychology	$13.95	_____
___ Opportunities in Teaching	$13.95	_____
___ Opportunities in Telecommunications	$13.95	_____
___ Opportunities in Television & Video	$13.95	_____
___ Opportunities in Veterinary Medicine	$13.95	_____
___ Opportunities in Waste Management	$13.95	_____
___ Outdoor Careers	$14.95	_____
___ Public Relations Career Directory	$17.95	_____
___ Radio and Television Career Directory	$17.95	_____
___ Travel and Hospitality Career Directory	$17.95	_____

INTERNATIONAL AND OVERSEAS JOBS

___ Almanac of International Jobs and Careers	$14.95	_____
___ Building an Import/Export Business	$14.95	_____
___ Complete Guide To International Jobs & Careers	$13.95	_____
___ Directory of Jobs and Careers Abroad	$14.95	_____
___ Directory of Overseas Summer Jobs	$14.95	_____
___ Getting Your Job in the Middle East	$19.95	_____
___ Guide To Careers in World Affairs	$13.95	_____
___ How To Get a Job in Europe	$17.95	_____
___ How To Get a Job in the Pacific Rim	$17.95	_____
___ How To Get a Job With a Cruise Line	$12.95	_____
___ International Consultant	$22.95	_____
___ International Directory of Voluntary Work	$13.95	_____
___ International Jobs	$12.95	_____
___ Job Hunter's Guide To Japan	$12.95	_____
___ Jobs For People Who Love Travel	$12.95	_____
___ Jobs in Paradise	$11.95	_____
___ Passport to Overseas Employment	$14.95	_____
___ Teaching English Abroad	$13.95	_____
___ Work Your Way Around the World	$17.95	_____
___ Work, Study, Travel Abroad	$16.95	_____

GOVERNMENT AND PUBLIC-ORIENTED CAREERS

___ The 171 Reference Book	$18.95	_____
___ ACWA: Administrative Careers With America	$15.00	_____
___ Almanac of American Government Jobs and Careers	$14.95	_____
___ Book of $16,000-$60,000 Post Office Jobs	$14.95	_____
___ Book of U.S. Government Jobs	$15.95	_____
___ Civil Service Handbook	$9.95	_____
___ Complete Guide To Public Employment	$15.95	_____
___ Complete Guide To U.S. Civil Service Jobs	$9.95	_____
___ Federal Jobs For College Graduates	$14.95	_____
___ Federal Jobs in Law Enforcement	$15.95	_____
___ Find a Federal Job Fast!	$9.95	_____
___ General Test Practice For 101 U.S. Jobs	$9.95	_____
___ Government Job Finder	$14.95	_____
___ How To Get A Federal Job	$15.00	_____
___ Law Enforcement Careers	$12.95	_____
___ Law Enforcement Employment	$19.95	_____
___ Paralegal	$10.95	_____
___ Promote Yourself (With the Federal Government)	$17.95	_____
___ Right SF-171 Writer	$17.95	_____

NONPROFIT CAREERS

___ Good Works	$18.00	_____
___ Great Careers	$36.00	_____
___ Non-Profits' Job Finder	$14.95	_____
___ Profitable Careers in Nonprofits	$14.95	_____

COMPUTER SOFTWARE

___ Computerized Career Assessment and Planning Program	$489.95	_____
___ Computerized Career Information System	$309.95	_____
___ EZ—D.O.T.	$299.95	_____
___ JOBHUNT Quick and Easy Employer Contacts	$49.95	_____
___ INSTANT Job Hunting Letters	$39.95	_____
___ Occupational Outlook on Computer	$129.95	_____
___ Perfect Resume Computer Kit (Personal)	$49.95	_____
___ Quick and Easy 171's (Individual)	$49.95	_____
___ ResumeMaker	$49.95	_____

VIDEOS

___ Find the Job You Want...and Get It! (4 videos)	$229.95	_____
___ How To Present a Professional Image (2 videos)	$149.95	_____
___ How To Set and Achieve Goals (2 videos)	$149.95	_____
___ Insider's Guide To Competitive Interviewing	$49.95	_____
___ Juggling Your Work and Family	$79.95	_____
___ Networking Your Way To Success	$79.95	_____
___ Self-Esteem and Peak Performance (2 videos)	$149.95	_____
___ Winning At Job Hunting in the 90s	$89.95	_____

Career Resources 117

JOB LISTINGS AND VACANCY ANNOUNCEMENTS

___ Federal Career Opportunities (6 biweekly issues)	$38.00	_____
___ International Employment Gazette (6 biweekly issues)	$35.00	_____

SKILLS, TESTING, SELF-ASSESSMENT

___ Career Exploration Inventory	$24.95	_____
___ Career Sort Assessment Instruments	$27.95	_____
___ Discover the Best Jobs For You!	$11.95	_____
___ Do What You Are	$14.95	_____
___ New Quick Job Hunting Map	$3.95	_____
___ Test Your I.Q.	$6.95	_____
___ Three Boxes of Life	$14.95	_____
___ Truth About You	$11.95	_____
___ What Color Is Your Parachute?	$14.95	_____
___ Where Do I Go From Here With My Life?	$10.95	_____
___ Wishcraft	$9.95	_____

EMPOWERMENT, SELF-ESTEEM, MANAGING CHANGE

___ 7 Habits of Highly Effective People	$11.00	_____
___ Bouncing Back	$14.95	_____
___ Courage To Fail	$18.95	_____
___ Do What You Love, the Money Will Follow	$8.95	_____
___ Dreams That Can Change Your Life	$18.95	_____
___ Softpower	$10.95	_____
___ Staying Up When Your Job Pulls You Down	$10.95	_____
___ Work With Passion	$9.95	_____
___ Your Own Worst Enemy	$19.95	_____

RESUMES, LETTERS, NETWORKING

___ Dynamite Cover Letters	$9.95	_____
___ Dynamite Resumes	$9.95	_____
___ Encyclopedia of Job-Winning Resumes	$16.95	_____
___ Great Connections	$11.95	_____
___ High Impact Resumes and Letters	$12.95	_____
___ How To Work a Room	$9.95	_____
___ Job Search Letters That Get Results	$12.95	_____
___ Network Your Way To Job and Career Success	$11.95	_____
___ No-Pain Resume Book	$14.95	_____
___ Perfect Cover Letter	$9.95	_____
___ Perfect Resume	$10.95	_____
___ Perfect Resume Strategies	$12.50	_____
___ Power Networking	$12.95	_____
___ Power Resumes	$12.95	_____
___ Resume Catalog	$15.95	_____
___ Resumes For High School Graduates	$9.95	_____
___ Resumes, Resumes, Resumes	$8.95	_____
___ Revising Your Resume	$13.95	_____

___ Smart Woman's Guide To Resumes
    and Job Hunting $8.95 _____
___ Sure-Hire Resumes $14.95 _____
___ Your First Resume $10.95 _____

DRESS, APPEARANCE, IMAGE, ETIQUETTE

___ Dressing Smart $19.95 _____
___ John Molloy's New Dress For Success $10.95 _____
___ Professional Image $10.95 _____
___ Professional Presence $21.95 _____

INTERVIEWS AND SALARY NEGOTIATIONS

___ Dynamite Answers To Interview Questions $9.95 _____
___ Interview For Success $11.95 _____
___ Listening: The Forgotten Skill $12.95 _____
___ Perfect Follow-Up Method To Win the Job $9.95 _____
___ Perfect Interview $17.95 _____
___ Power Interviews $12.95 _____
___ Salary Success $11.95 _____
___ Sweaty Palms $9.95 _____

WOMEN AND SPOUSES

___ Balancing Career and Family $7.95 _____
___ Congratulations: You've Been Fired! $8.95 _____
___ Delights, Dilemmas, and Decisions $18.95 _____
___ Female Advantage $19.95 _____
___ Hardball For Women $21.95 _____
___ New Relocating Spouse's Guide To Employment $14.95 _____
___ Resumes For Re-Entry: A Handbook For Women $10.95 _____
___ Rights of Passage $21.95 _____
___ Smart Woman's Guide To Resumes and Job Hunting $8.95 _____
___ Survival Guide For Women $16.95 _____
___ Women's Job Search Handbook $12.95 _____

MINORITIES AND DISABLED

___ Directory of Special Programs For
    Minority Group Members $31.95 _____
___ Job Hunting For People With Disabilities $14.95 _____
___ Minority Organizations $49.95 _____
___ Work, Sister, Work $18.95 _____
___ Yes You Can $12.95 _____

COLLEGE STUDENTS

___ College Majors and Careers $15.95 _____
___ Complete Resume and Job Search Book
    For College Students $9.95 _____
___ Graduating To the 9-5 World $11.95 _____

___ Liberal Arts Jobs $10.95 _____
___ MBA's Guide To Career Planning $16.95 _____

ENTREPRENEURSHIP AND SELF-EMPLOYMENT

___ 101 Best Businesses To Start $15.00 _____
___ 164 More Businesses Anyone Can Start $12.95 _____
___ 184 Businesses Anyone Can Start $12.95 _____
___ Best Home-Based Businesses For the 90s $10.95 _____
___ Entrepreneur's Guide To Starting a Successful Business $16.95 _____
___ Starting On a Shoestring $14.95 _____

                  SUBTOTAL   _____

                  Virginia residents add 4½% sales tax   _____

                  POSTAGE/HANDLING ($3.00 for first
                  title and 75¢ for each additional book) $3.00

                  Number of additional titles x 75¢ ------------   _____

                  TOTAL ENCLOSED ----------------   _____

SHIP TO:

NAME _____

ADDRESS _____

_____

_____

    [ ] I enclose check/moneyorder for $ _____ made
        payable to IMPACT PUBLICATIONS.

    [ ] Please charge $ _____ to my credit card:

        Card # _____

        Expiration date: _____ / _____

        Signature _____

"JOBS AND CAREERS FOR THE 1990s" CATALOG

To receive your free copy of *"Jobs and Careers For the 1990s,"* complete the following form or send your name and address to:

IMPACT PUBLICATIONS
ATTN: Free Catalog
9104-N Manassas Drive
Manassas Park, VA 22111

NAME _____

ADDRESS _____

_____

_____